LIFTOFF

ELON MUSK AND

THE DESPERATE

EARLY DAYS

THAT LAUNCHED

SPACEX

ERIC BERGER

WILLIAM MORROW
An Imprint of HarperCollinsPublishers

LIFTOFF

HarperCollins books may be purchased for educational, business, or sales promotional use. For information, please email the Special Markets Department at SPsales@harpercollins.com.

FIRST EDITION

Designed by Elina Cohen
Photos from Shutterstock by Dima Zel and ixpert

Library of Congress Cataloging-in-Publication Data has been applied for.

ISBN 978-0-06-297997-1

21 22 23 24 25   LSC   10 9 8 7 6 5 4 3 2 1

CONTENTS

LIFTOFF

A fat, red sun sank into the Texas horizon as Elon Musk bounded toward a silvery spaceship. Reaching its concrete landing pad, Musk marveled up at the stainless steel, steampunk contraption looming above, which shone brilliantly in the dying light. "It's like something out of a *Mad Max* movie," he gushed about the first prototype of his Mars rocket, nicknamed Starhopper.

Musk traveled to his South Texas rocket factory in mid-September 2019 to track progress of SpaceX's Starship vehicle, the culmination of nearly two decades of effort to move humans from Earth to Mars. Weeks earlier, Starhopper soared into the clear skies above the coastal scrubland, located just this side of the Mexico border. And then, it very nearly crashed. Luckily, the Federal Aviation Administration had restricted the flight's maximum altitude to five hundred feet, so when engineers lost control during Starhopper's descent its landing legs merely crushed through the pad's steel-reinforced concrete, rather than erupting into a ball of flame. Musk laughed at this thought. For much of SpaceX's lifetime he has fought against regulators, always seeking to go faster, to push higher. "This time," he quipped, "the FAA saved us."

This was his first visit to Starhopper since. Musk made the rounds, high-fiving a handful of employees and enjoying the moment with three of his sons who had come along for the weekend trip from Los Angeles. Starhopper, he explained to the boys, is made from stainless steel, the same stuff in pots and pans.

This stainless steel, however, had the look of being left on a stovetop's open flame for too long. The evening's deepening darkness could not mask extensive charring on the metal. Standing beneath Starhopper, Musk peered upward into the cavern housing a large fuel tank that had fed propellant to a Raptor rocket engine. "It's in remarkably good shape considering we had an inferno in there," he said.

Elon Musk traveled a long road to reach these plains rolling down to the Gulf of Mexico. In 2002, Musk founded SpaceX with the intention of eventually building spaceships that would take hundreds, and then thousands, of human settlers to Mars. Though a cold, likely dead, and nearly airless world, Mars nonetheless offers humanity the best place to expand beyond Earth. Mars has polar ice caps, useful chemicals in its thin atmosphere, and material to scratch out a living. It also is relatively close, as planets go.

Over the years, Musk has accomplished a number of remarkable feats with SpaceX, flying astronauts into space, landing rockets on boats, and remaking the global aerospace industry. But those achievements pale next to the audacity of trying to send humans to Mars, which remains far beyond the present-day capability of NASA or any other space agency around the world. Even with an annual budget approaching $25 billion a year, and some of the smartest scientists and engineers anywhere, the space agency that landed humans on the Moon remains several giant leaps away from sending a few astronauts to Mars.

Musk wants to build a city there. Perhaps it is better to say something inside Musk relentlessly drives him to do this. He long ago decided that for humanity to have a long-term future it must expand to other worlds, with Mars offering the best place to start. This is extremely hard because space is an insanely dangerous place, permeated by radiation, and with certain death always lurking on the other side of thin, pressurized walls. The amount of water, food, fuel, and clothing needed to sustain a months-long outbound mission to Mars is astounding, and once there people must actually have somewhere to survive on the surface. The largest object NASA has ever sent to the surface of Mars, the Perseverance rover, weighs about one ton. A single, small human mission would probably require fifty times the mass. For a sustainable human settlement, Musk thinks he probably needs to ship 1 million tons to Mars. This is why he is building the massive, reusable Starship vehicle in Texas.

In many ways, SpaceX is vastly different today from the company Musk started long ago. But in important ways, it remains exactly the same. With the Starship project, SpaceX has returned to its earliest, scrappy days when it strove to build the Falcon 1 rocket against all odds. Then, as now, Musk pushed his employees relentlessly to move fast, to innovate, to test, and to fly. The DNA of the earliest days, of the Falcon 1 rocket, lives on in South Texas today at the Starship factory. And a huge photo of a Falcon 1 launch hangs on the wall of Musk's personal conference room at the company's headquarters in California.

To understand SpaceX, where it aspires to go, and why it just might succeed, one must voyage back to the Falcon 1 rocket and dig up the roots. The seeds for everything SpaceX has grown into today were planted during the early days of the Falcon 1 program by Musk. Back then he sought to build the world's first low-cost, orbital rocket. All of the aspirational talk about Mars would mean nothing if SpaceX could not put a relatively simple rocket like the Falcon 1 into orbit. And so, with a burning intensity, he pressed toward that goal. SpaceX began with nothing but an empty factory and a handful of employees. This small group launched its first rocket less than four years later and reached orbit in six. The story of how SpaceX survived those lean, early years is a remarkable one. Many of the same people who made the Falcon 1 go remain at SpaceX today. Some have moved on. But all have stories about those early, formative years that remain mostly untold.

The men and women who helped Musk bring SpaceX through its darkest days hailed from farm country in California, from the suburbs of the Midwest, from East Coast cities, from Lebanon, Turkey, and Germany. Musk hired them all, molded them into a team, and coaxed them to do the nearly impossible. Their path to orbit led from the United States to a small tropical island about as far from a continental landmass as one can get on this world. And out there in the middle of the Pacific Ocean, the company very nearly died multiple times.

More than a decade later Musk and SpaceX have traversed the chasm separating failure and success. After perusing Starhopper at sunset, he spent several hours touring his rocket shipyard in South Texas. Through the night, as a full moon rose, employees banged and welded and hefted a full-sized Starship prototype from rolls of stainless steel. The hour had reached near midnight when he and his boys emerged from a construction trailer. As his kids tumbled into the waiting black SUV, Musk paused to look up at the towering Starship under construction. It appeared as much a skyscraper as a spaceship.

Taking it all in, a childlike smile broke out over his face. "Hey," Musk said, turning to me. "Can you believe that thing, or something like it, is going to take people to another planet for the first time in 4.5 billion years? I mean, probably. It may not work. But it probably will."

EARLY YEARS

September 2000–December 2004

For those so bold as to dare fly to Mars, the summer of 2003 offered a hopeful sign of things to come. Due to the quirks of planetary motion, in July the red planet made its closest approach to Earth in sixty thousand years. At the time, a small company named SpaceX had only just begun to cut metal on its first rocket. Although its inaugural launch remained a few years away, the firm's founder, Elon Musk, had already taken the first step toward Mars. He understood he would go nowhere without the right people. So interview by interview, Musk sought out the brilliant and creative engineers who would commit themselves wholly to his goal—and make the impossible possible. He was beginning to find them.

Brian Bjelde was oblivious to Mars's close approach and Musk's dreams that summer when he received a phone call from a former college classmate. They had bonded during late nights in the University of Southern California's aerospace lab, tinkering with vacuum chambers and small satellites. The friend, Phil Kassouf, spoke rapturously about his new job working for a hard-charging multimillionaire from Silicon Valley. The guy had crazy plans to build a rocket and one day travel to

Mars. You should come by for a tour, Kassouf said, and gave his friend an address near the Los Angeles airport.

Bjelde was living a charmed existence at the time. The cherubic twenty-three-year-old had risen from modest means in California's rural farm country to make good in the big city. After graduating from U.S.C. as an aerospace engineer, Bjelde took a job at NASA's prestigious Jet Propulsion Laboratory, just north of Los Angeles. In turn, NASA paid for graduate school at U.S.C. As an advisor to a fraternity, Bjelde enjoyed free housing along with his pick of the best weekend parties.

So when Bjelde rolled up to SpaceX's modest headquarters in El Segundo, he really had just come for the tour. "You walk in, and there's a desk, and there's these two double glass doors," Bjelde said. "I walked through the office, shaking hands. There were gray cubicles. There was really nothing on the tour. Only an empty factory. They had just glossed off the factory floors."

What struck Bjelde most of all was the Coke machine in the break room. Musk had imported this innovation from Silicon Valley—unlimited free soda, to keep the workforce caffeinated at all hours. For someone from academia, and the sober environment at NASA, this was a novelty. As he moved through the office, one of the dozen or so people in the cube farm asked Bjelde about his projects at the Jet Propulsion Laboratory, which builds robotic spacecraft to explore the Solar System. Bjelde explained about his use of semiconductors, plasma etching, and vapor pressure to develop new propulsion techniques for small satellites.

Sure, someone responded, but what did he think about propulsion for big systems? Like, say, rockets? Suddenly, it clicked. Bjelde had not really been invited for a tour and as many Cokes as he could drink. This was a job interview.

"I ended up in this room," he said. "Unbeknownst to me, it was called the meat locker because it was so cold. Somehow, in the HVAC circuit, it got the super flow. It was freezing in there."

Various people rotated through. His friend, Kassouf, came first. Then Phil's boss, the company's vice president of avionics, Hans Koenigsmann, spoke with Bjelde. Eventually, Musk himself walked in. Only a decade older than Bjelde, Musk already was a very wealthy, increasingly famous entrepreneur. To break the ice, Bjelde made the usual small talk—it's nice to meet you, I've heard a lot about you, I'm excited to be here. The hyperobservant Musk, never one much for pleasantries, moved straight into questions.

"Do you dye your hair?" Musk asked.

Somewhat flustered, Bjelde replied that he did not. One of Musk's common tactics during an interview involves throwing a person off-kilter, to see how a potential employee reacts. In Bjelde, however, he had found someone with the gift of gab. Bjelde can talk to anyone. So after quickly recovering, he asked Musk, "Is this an icebreaker? Because it's working."

But Musk said he was serious. He had noticed that Bjelde's eyebrows were very light, and his hair darker. The young engineer explained that the disparity was natural. Soon, they were laughing.

During the thirty-minute interview Musk probed into Bjelde's background, but also shared his vision for SpaceX, founded to make humanity a truly spacefaring civilization. The success of NASA's Apollo Moon program in the 1960s had spurred a wave of student interest in math and science, and led to a generation of engineers, scientists, and teachers. But this tide had ebbed by the turn of the century. Bjelde's generation had grown up with the space shuttle, and its endless revolutions around Earth in low-Earth orbit, not the derring-do of the Apollo explorers. Unlike Bjelde, who had chosen his major literally because aerospace was listed first alphabetically under engineering, most of the cool kids were not doing space anymore. They were into medicine, investment banking, or tech.

Musk had been among those leading the digital revolution. With

PayPal he had helped take the banking industry online. And everywhere from communications to health care, the digital transformation had begun accelerating. Yet the stodgy aerospace industry seemed to be going backward. Companies in the United States and Russia still used the same decades-old technology to launch rockets into space, and the price kept going up. It seemed like things were going in the wrong direction, so Musk had founded SpaceX, and now a year later he sought to move from basic designs into developing hardware. Musk wanted Bjelde to help with the rocket's electronics.

It was a lot for Bjelde, sitting in that frigid room, to take in. He had a comfortable government job, a promising academic career, and an active social life. SpaceX would strip all of that away. From talking to Kassouf about SpaceX's intense environment, Bjelde knew coming to work for Musk would turn his life upside down. And Musk could offer no guarantees of success. How could such a small team build a rocket capable of reaching orbit, anyway? No privately funded company had ever succeeded at something like this before, and many had failed trying. After his interviews, Bjelde wondered if he'd been fed mostly empty promises.

A few days later, he received an email from Musk's assistant, Mary Beth Brown, at one in the morning. Did he want a job? Bjelde realized this company operated at its own speed.

At first, Bjelde tried to negotiate for a higher salary. NASA paid him a comfortable $60,000 a year, along with his tuition. SpaceX offered less. For a chance to work with a visionary, on an inspiring project with a mission he could embrace, Bjelde would have to eat a salary cut. In thinking it over, he recalled a high school chemistry teacher named Ms. Wild, who had an eccentric bucket list. As a student, Bjelde saw her embrace opportunities when the chance arose, ticking off items such as belly dancing at the foot of the Egyptian pyramids. So this offer appealed to Bjelde and his sense of adventure, and he decided to seize this chance with Musk. After all, getting to Mars was a crazy hard goal. Nearly impossible. But *not* impossible.

"I'd love to think that we could live in a world where in our lifetime, during this short little blink of an eye where we get to be here, that we can make a rapid change to where you or I, or anyone, could have the means to afford it," he said of traveling to Mars. "That's something that's right in front of us. It's within our reach."

Later, Bjelde learned that before his visit to SpaceX, Kassouf had gone to bat for him. The company needed someone who could build electronics for a rocket's brains, the hardware and software to help the booster fly straight. Bjelde wasn't even an electronics engineer. But Kassouf had told Musk about the long hours they'd worked together at U.S.C., the all-nighters, and his friend's passion for solving hard problems. Kassouf had effectively put his badge on the table for his buddy— yes, Bjelde would lay it all on the line for SpaceX and the Falcon 1 rocket. In August 2003, Brian Bjelde, funny-colored eyebrows and all, officially became employee number fourteen at SpaceX.

The story of SpaceX begins toward the end of the year 2000, on the other side of the United States. Elon Musk was driving on the Long Island Expressway with a friend and fellow entrepreneur named Adeo Ressi, shortly after PayPal's board of directors had ousted Musk as chief executive. Not yet even thirty years old, Musk had come a long way in a short while. Since arriving in the United States less than a decade earlier, he had earned Ivy League degrees in economics and physics, and founded two wildly successful companies. So what, Ressi wanted to know, did he plan to do next?

"I told Adeo I had always been interested in space, but I didn't think that was something a private individual could do anything about," Musk said. Three decades had passed since Apollo's heyday. Surely, he thought,

NASA must be well on its way to Mars. Later that day, Musk, still thinking about the conversation, checked out NASA's website. To his surprise, he could not find any plans for sending humans to Mars. Perhaps, he thought, the site was just poorly designed.

But no: NASA had no such plans, as Musk soon discovered when he began attending space conferences back in California. Private groups were beginning to do some interesting things, however. He got involved with ventures like the Planetary Society's first project to develop a solar sail. The member-funded organization was building a reflective sail that would unfurl in space and be propelled by momentum from solar photons. Musk also supported the XPRIZE Foundation, which offered $10 million to the first group to build a private spaceship that could take people on short, suborbital flights. Later in 2001, Musk devised a private space plan of his own to inspire public support for NASA and the exploration of Mars. Musk sought to build a small biosphere and launch it to the red planet. He called it Mars Oasis.

"The idea was to grab some Martian soil, and bring that into the growth chamber," said Chris Thompson, an aerospace engineer for Boeing who helped Musk with concepts for the small Martian lander. "We would mix it with some soil from Earth, drop some seeds in, and have a webcam broadcast the plant growth back to Earth."

As Thompson and a few other engineers worked on the payload side of the biosphere project, Musk and his advisors twice traveled to Russia to try and buy a refurbished intercontinental ballistic missile for the mission. The Russians had no respect for Musk, seeing him as a dilettante, and so they offered him outrageously high prices for their old boosters. And Musk feared that if he agreed to their price, they would only raise it after he wrote the initial check. "The last trip from Russia, I was like, man, the price just keeps going up, it doesn't feel like this project is going to be successful," Musk said. "I wondered what it would take to build our own rocket."

One of his advisors, an engineer and aspiring businessman named Jim Cantrell, urged him to think seriously about doing just that. So Musk began to meet with rocket scientists in the Los Angeles community, a hub for aerospace engineers. Soon, he picked up other advisors for his fledgling effort, including John Garvey, who had worked with Thompson at Boeing, and later a rising star in rocket engines, Tom Mueller. Numerous other entrepreneurs had tried playing at rocket science before, Musk well knew. He wanted to learn from their mistakes so as not to repeat them.

In February 2002, Garvey arranged for Musk to visit the launch site for the Reaction Research Society, a famed rocketry club in Southern California. The millionaire had come ill prepared for the brisk winds and chilly temperatures of the Mojave high desert. "I think at the time it was probably eighteen degrees out," Thompson said. "And he shows up in slacks and Neiman Marcus shoes and a skinny leather jacket." But Musk asked good questions and listened intently. He had been reading everything he could get his hands on about rockets, from old Soviet technical manuals to John Drury Clark's iconic book on propellants, *Ignition!*

As he learned more about rockets, Musk also gained a deeper understanding of the deficiencies of the U.S. launch industry. His vision for Mars Oasis had been to inspire the public, leading to greater funding for NASA, and ultimately extending humanity's reach to the Moon and Mars by continuing the legacy of Apollo. He saw the problems with NASA and the global launch industry were more systemic than mere funding. Even if Mars Oasis succeeded, and NASA's budget doubled, he realized it would probably only lead to more flags-and-footprints missions. Musk wanted nothing less than human expansion into the Solar System, and the settlement of its worlds.

"I began to understand why things were so expensive," he said. "I looked at the horses that NASA had in the stable. And with horses like

Boeing and Lockheed, you're screwed. Those horses are lame. I knew Mars Oasis would not be enough."

The first step toward solving the multiplanetary problem, then, was bringing down the cost of the launch. If NASA and private companies spent less money getting satellites and people into space, they could do more things in space. And more commerce would open still more opportunities. This awakening galvanized Musk into action.

That spring Musk called a meeting of about fifteen or twenty prominent aerospace engineers at the Renaissance Hotel at the Los Angeles airport. Many had come at the behest of Mike Griffin, a leader in the community who would become NASA's administrator three years later, and whom Musk had relied upon for advice. Garvey, Mueller, and Thompson also had seats at the table.

"In typical Elon fashion, he kind of showed up a little bit late, which clearly annoyed a lot of the older-guard aerospace executives that were in the room," Thompson said. "He walks in and basically announces that he wants to start his own rocket company. And I do remember a lot of chuckling, some laughter, people saying things like, 'Save your money kid, and go sit on the beach.'"

The kid was not amused. If anything, the doubts expressed at this meeting, and by some of his confidants, energized him more. Several friends had already tried to dissuade Musk from this venture. Ressi created an hourlong video compilation of rocket failures and forced Musk to sit down and watch it. Peter Diamandis, an engineer, told Musk of all the other entrepreneurs who had tried, and failed. "He talked my ear off, and said I would lose all my money," Musk said.

As he looked around the table during the meeting at the Renaissance Hotel, therefore, Musk searched among the doubters to find the few believers. Musk wanted people who embraced a challenge rather than shrank from it, optimists rather than pessimists. In April, Musk offered five people the opportunity to be a member of the company's "founding team." He had cleared about $180 million from PayPal, and figured he could risk half of that on a rocket company, and still have plenty left over. Musk brought the cash, and wanted his early employees to invest in sweat equity.

Only two of the five accepted. Offered the title of chief engineer, Griffin said he preferred to stay on the East Coast near Washington, D.C., where he was an important player in national space policy. Musk nixed the idea of a cross-country commute, and this was probably for the best. Though brilliant, Griffin possessed a headstrong personality similar to Musk, and they would have clashed. Musk continued hunting for the right person, but, he said, "Nobody who seemed to be good would join, and there was no point in hiring somebody who wasn't good." Elon Musk assumed the role of chief engineer himself.

He also liked Cantrell, thinking the smooth-talking engineer could serve as the chief of business development for SpaceX. But Cantrell did not want to relocate, either. To move from Utah, he asked for a large salary and all sorts of guarantees. "He ultimately decided not to join," Musk said. "He was only ever a consultant for a brief period of time."

A third no came from John Garvey, something of a surprise given that the rocket scientist had been an enthusiastic supporter of the venture. Garvey thought Musk's concept for a rocket that could lift one thousand pounds into space might be too ambitious and preferred a lighter design. He also wanted Musk to buy out his small aerospace company, Garvey Spacecraft Corporation. And, Musk said, Garvey wanted a lofty title— chief financial officer. This baffled Musk, as Garvey had no background in finance.

After three rejections, Musk had just two people left on his list. Mueller had seen rocket entrepreneurs with good plans and no money before, and bad plans and plenty of money. In Musk, he found someone both with ideas he liked and with enough capital to see the venture through the difficult design-and-development phase. Above all else, Mueller relished the challenge of building a new rocket engine all his own. When Musk offered him the chance to do that, along with shares in the company, Mueller talked it over with his wife. He had a stable job at a large aerospace company. But his wife knew he would rue passing on this opportunity. She encouraged him to take the job. Mueller did. As the first to sign, Mueller became employee number one on SpaceX's payroll.

Thompson, with a young family, shared the same hesitation about leaving a comfortable position in the aerospace industry. During a phone call in late April, Musk sought to allay those concerns. Musk recognized what Thompson and Mueller were walking away from, so he put two years' worth of salary for both engineers into an escrow account. That way, if Musk decided to prematurely pull the plug on the venture, they would still have a guaranteed income. This helped Thompson convince his wife he should take the job. His only regret? That he mulled it over long enough to get tagged as employee number two.

On May 6, 2002, Musk founded Space Exploration Technologies. Originally, he, Mueller, and Thompson referred to the company as S.E.T. After a few months, Musk came up with a catchier nickname—SpaceX.

Initially, the trio continued to meet in airport hotels. Mueller would report on his efforts to design a new rocket engine for the booster, which Musk would soon christen the Falcon 1. The name came from the iconic spaceship in *Star Wars*, and because the rocket would have a single main engine. As vice president of propulsion, Mueller had to develop this engine, the rocket's fuel tanks, and the plumbing that carried chilled liquid propellants throughout. Thompson, vice president of "structures," would design the lightest possible frame made of aluminum alloy, and the mechanisms by which the rocket separated during flight.

The company still needed someone to oversee avionics, the Falcon 1's onboard computer and software. Had Garvey hired on, this job probably would have fallen to him. In his place, Thompson recommended a German engineer named Hans Koenigsmann, who worked at a small southern California aerospace firm named Microcosm. Musk had previously met Koenigsmann on that cold day in Mojave a few months earlier, and the German engineer immediately bought into Musk's plan for building a low-cost rocket with a small team.

"Here's the thing," Koenigsmann said. "I didn't want to be an astronaut. That's not my thing. But what did intrigue me was trying to build a rocket with two hundred people instead of twenty thousand. To almost build this in a garage. Can I use a computer I can buy for $500, versus one I can buy for $5 million? It seemed to me that's what he wanted to do."

This was precisely what Musk wanted to do. Because they were spending his money, Musk gave employees an incentive to be frugal with it. Although Musk retained a majority of shares, early hires received large chunks of stock. When an employee saved the company $100,000 by building a part in-house instead of ordering one from a traditional supplier, everyone benefited.

With his core team in place, Musk moved his company into a large, white building at 1310 East Grand Avenue in El Segundo. The thirty-thousand-square-foot facility seemed cavernous at the time, with fewer than a dozen employees seated in a central office area, and an empty factory out back. Over time, the company would fill it up, expanding like kudzu into surrounding office buildings. But in those earliest years, SpaceX had only a few cubicles, a few computers, and almost no organization.

After his staid government job at NASA, Bjelde felt the culture shock immediately upon starting work. Before he could log on to a computer at NASA, Bjelde had undergone a detailed security screening process and multiple orientations. To operate machines that would steer electron beams, Bjelde had sat through days of training courses.

"At SpaceX back then, there was none of that," Bjelde said of his first day on the job. "You show up. The door is not locked. There's no one at the front desk. I met Hans, and he gave me a packet that had some materials about benefits and things like that. And then he told me what I needed to do." Orientation done.

In the folder, Bjelde also found a few rudimentary documents someone had cobbled together about the Falcon 1's flight termination system. Every rocket launching from U.S. soil must have a mechanism that allows the operator of a launch range, typically the U.S. Air Force or Army, to radio a destruct signal to a booster if it veers off course after liftoff. This system must be fail-safe, because an errant rocket might potentially threaten populated areas. Numerous government agencies had to sign off on its design. So first, Bjelde had to learn how to build the system. Then he had to design it, and obtain all the necessary paperwork from the government, before finally building and testing it. And he had to hurry because Musk wanted to launch in a year.

They spent their long and often intense days together in close confines. Musk kept a mostly laissez faire attitude toward his workplace. He offered just a few hard and fast rules: no strong smells, no flickering lights, and no loud noises in the cubicle farm they all shared. Often, they worked until well after midnight. Bjelde, slumped under his desk, recalls being kicked awake on more than one occasion to help finish writing a proposal.

Their close and nearly continual proximity led to easy collaboration. The team was so small that everybody knew everybody, and each employee pitched in as needed with other departments.

"Everybody was expected to carry their own weight, and then a bunch more," Thompson said. "If Mueller needed help with something, I would stop, drop, and roll and help Tom. If help was needed designing a test stand, I would step up. Or if I needed help, someone would just jump right in. It was definitely multiple hats, up to and including janitor."

Indeed, during the early years of SpaceX the company had no real support staff beyond Musk's all-star assistant, Mary Beth Brown. This included a lack of custodial staff. After she hired on in August 2002 as head of sales, Gwynne Shotwell remembers organizing a meeting with government customers for a potential satellite launch. She checked in on the company's upstairs conference room to make sure it was suitable for the military brass. "They were going to be there in an hour, and it was a mess," she said. "So I got out the vacuum. The vice president of sales vacuuming, and then trying to figure out coffee."

Each employee took his or her turn on Friday ice cream runs, too. After a Cold Stone Creamery opened less than a mile away, an office tradition quickly developed. An email would go around with an order sheet, and each employee would write their name or nickname, and their flavor of choice. "Rat Chicken"—Bjelde—might order Birthday Cake ice cream. Then someone, a new hire this week or a vice president the next, would take the only SpaceX company credit card down to the store, place the orders, and return to the company's offices.

"No job was beneath us," Bjelde said.

The growing team also bonded over computer games. Following a long day of work, most of the employees in the office would put the phones on their desks into conference mode. The office would come alive with banter and bravado as the SpaceX employees loaded the computer game *Quake III Arena*, a first-person shooter that allowed multiple players to join, and battle one another in death matches. Each participant would choose a playable character and a weapon, and look for targets on the virtual playing field.

"There were days where we played that game until three o'clock in the morning," Thompson said. "We'd be screaming and yelling at each other like a bunch of lunatics. And Elon was right there in the thick of it with us."

Not everyone played late into the night. Asked about the *Quake* parties,

Shotwell, one of the few women in the office at the time, responded with a laugh. "No, I was never part of that," she said. "It was good work time for me." Sometimes, Shotwell said, she and Mary Beth Brown would joke that maybe they should get a game of *My Little Pony* going.

In truth, the hard-working team needed the escapism of computer games. It felt cathartic to frag the boss, who so often demanded the impossible over the course of eighty-hour weeks. "We sometimes joke that SpaceX is like dog years," Bjelde said. "You get like seven years in one. And it's true."

Creating a rocket company involved a lot of travel. SpaceX needed to find a test site for its engines and fuel tanks, and then a location to launch from. Musk had to meet with potential customers. And he and his vice presidents had to find suppliers for key parts of the Falcon 1 that could not be built in-house. Although Musk wanted to develop the rocket's engine at SpaceX, he was willing to buy pressurized tanks from suppliers. Tanks are not simple, as they must both be lightweight as well as capable of storing extremely cold and combustible fuels at high pressure.

In late 2002 Musk arranged a meeting with a tank manufacturing company in Green Bay, Wisconsin. He and a handful of engineers arrived the night before, staying at a Holiday Inn Express. Chris Thompson and another early employee named Steve Johnson, seeking to impress Musk, woke early so they would already be eating at the small breakfast buffet when the boss arrived.

"Elon comes down, and he walks over to the breakfast bar and he picks up a package of Pop-Tarts," Thompson said. "And the funniest thing to me was the fact that most of us take Pop-Tarts for granted. He was transfixed. This was like a scene out of *2001: A Space Odyssey*, when the apes examine the monolith. It was clearly the most fascinating thing he had seen that morning."

Eventually, Musk realized that Pop-Tarts were best enjoyed toasted. So he opened a package and put two of them into the toaster, Thompson said. Only Musk made the rookie mistake of inserting the pastries horizontally, rather than vertically. When they popped back up, he had to stick his fingers into the toaster to grab his breakfast. This was a problem, and at about six in the morning Musk proceeded to scream, at full volume, "Fuck, it burns! Fuck, it burns!" Two older ladies at the front desk, nearby, watched in mortified silence.

It worked out in the end. The Green Bay company they met with could not help SpaceX, but it suggested Spincraft, another manufacturer near Milwaukee. SpaceX had found its fuel tank supplier.

These kinds of trips, of which there were many, helped Musk bond with his senior leaders. He could be difficult to work for, certainly. But his early hires could immediately see the benefits of working for someone who wanted to get things done and often made decisions on the spot. When Musk decided that Spincraft could make good tanks for a fair price, that was it. No committees. No reports. Just, done.

This decisive style carried over into meetings back at the office in El Segundo. Musk would convene his different teams in a small conference room, be it his engineers working on propulsion, or structures, or avionics, and run down the major issues. If an engineer faced an intractable problem, Musk wanted a chance to solve it. He would suggest ideas and give his teams a day or two to troubleshoot, then report back to him. In the interim, if they needed guidance, they were told to email Musk directly, day or night. He typically responded within minutes. Over the course of a single meeting Musk could be, at turns, hilarious, deadly serious, penetrating, harsh, reflective, and a stickler for the finest details of rocket science. But most of all, he channeled a preternatural force to move things forward. Elon Musk just wants to get shit done.

The engineers sitting in those seats around the conference table had to possess a certain amount of mania, too. First they had to accept Musk's ambitious, if not all-but-impossible vision. But it takes a rarer breed still who can sprint through thickets of technical problems as someone urges them on, faster and faster. One of Musk's most valuable skills was his ability to determine whether someone would fit this mold. His people had to be brilliant. They had to be hardworking. And there could be no nonsense.

"There are a ton of phonies out there, and not many who are the real deal," Musk said of his approach to interviewing engineers. "I can usually tell within fifteen minutes, and I can for sure tell within a few days of working with them." Musk made hiring a priority. He personally met with every single person the company hired through the first three thousand employees. It required late nights and weekends, but he felt it important to get the right people for his company.

Take Phil Kassouf. Only weeks after Koenigsmann joined SpaceX, he needed to hire an electrical engineer to help design and fabricate printed circuit boards for the Falcon 1's onboard computer. The German had known Kassouf from his internship earlier that year at Microcosm. Not much fazed Kassouf, a precocious twenty-one-year-old used to hardship. He had grown up in war-torn Lebanon and left his family to come to the United States for college. He had brains, but little money. Lacking the means to attend MIT or Harvard, Kassouf took a full ride offered by U.S.C. He was not yet finished with his undergraduate studies when Koenigsmann urged him to visit the company's new offices in El Segundo.

Not long into his tour, Kassouf found himself sitting across from the entrepreneur with an intense stare and a penchant for throwing interview subjects off their game. As part of his interview process, Musk wanted not to test a person's knowledge but rather his or her ability to think. Musk's first question to Kassouf was therefore an engineering riddle.

"You're somewhere on Earth," Musk said. "You've got a flag and a compass. You plant the flag in the ground. You take a look at the compass, and you see it points south. So you walk a mile south. Then you turn, and you walk a mile east. Then you turn, and you walk a mile north. To your surprise, you're right back at the flag. Where are you?"

Kassouf thought it through. He could not be at the equator, because he'd walk a square. It could not be the South Pole, either, because of the compass reading. It must therefore be the North Pole, as the ninety-degree turns there end up forming three sides of a triangle at the top of the sphere. That was the correct answer. Musk began to move on to his next question, but Kassouf cut him off. "Wait, there's another place you could be."

Now Musk was interested.

"If you're north of the South Pole," Kassouf continued, "there's a place where the circumference of the Earth is exactly a mile. If you start a mile north of that, and go south one mile, go all the way around the Earth, and come back up a mile, you're in the same spot."

That was true, Musk acknowledged. Then he stopped asking Kassouf riddles and began discussing what Koenigsmann needed help with. It did not matter that Kassouf was just twenty-one, or that he lacked a college degree. He could do the job.

When Musk identified someone he wanted to hire, he could be relentless. In the spring of 2004, Bulent Altan had nearly finished a master's degree in aeronautics at Stanford. He planned to find work in the Bay Area where his wife, Rachel Searles, had already taken a desirable job at Google. However, a couple of Altan's engineering friends from graduate school had recently moved to Los Angeles to work at SpaceX. One of them, Steve Davis, texted Altan to say he would love the company and should come for a visit.

A native of Turkey who speaks with a distinct accent, Altan had moved to the United States only two years earlier. After studying computer science in Germany, he had found northern California to his liking. The thought of relocating so soon, especially to crowded, smoggy Los Angeles, held little appeal. So he made the trip intending just to see Davis and his other friends. Yet when he visited them at the El Segundo factory, Altan soon was swept up in SpaceX's mystique, as the company raced to complete its first flight-ready version of the Falcon 1 rocket. By the time he met with Musk, Altan realized he wanted to work there. But what of those Bay Area plans?

Davis had anticipated his friend's issue. Having convinced Musk they needed to bring the brilliant young engineer from Turkey on board, it became a matter of solving the problem. His wife had a job in San Francisco? She would need one in Los Angeles? "These were solvable problems, and Elon's better at solving problems than almost anyone else," Davis said.

Musk therefore came into his job interview with Altan prepared. About halfway through, Musk told Altan, "So I heard you don't want to move to L.A., and one of the reasons is that your wife works for Google. Well, I just talked to Larry, and they're going to transfer your wife down to L.A. So what are you going to do now?"

To solve this problem, Musk had called his friend Larry Page, the cofounder of Google. Altan sat in stunned silence for a moment. But then he replied, given all of that, he supposed he would come work for SpaceX.

The next day Searles went into her job at Google, and her manager said the oddest thing had just happened. Larry Page had called to say she could now work from the company's Los Angeles office if she wanted.

Relative to other aerospace companies, Musk had a lot to offer prospective employees. Florence Li had interned at both Boeing and NASA before interviewing for a full-time job at SpaceX. Musk made a compelling pitch for his vision of spaceflight. But more than that, he empowered his engineers. At SpaceX, new hires could rapidly grow their skills and take on new responsibilities. There was almost no management then, and everyone worked on the rocket. "A big thing was really having to learn to think, since nobody gave you a cookie-cutter job and told you what you do," Li said. "That really made us all much better engineers."

Kassouf sometimes called former classmates who had taken jobs in places like Austin, Texas, or Tucson, Arizona, and they would compare notes. One friend at Lockheed Martin worked on F-35 stealth aircraft, a lucrative program for the company. Eventually, the Air Force would buy more than two thousand units at a cost of $85 million each. It may sound like glamorous work, but it was not. Kassouf's friend had just a single job, finding a supplier for a bolt on the aircraft's landing gear and ensuring that it met all quality specifications. That single bolt was the totality of his employment. Although his friend did admit to boredom at work, he liked his house and his lifestyle away from the job. SpaceX offered the opposite experience. Work was thrilling and all encompassing. "It is hard to describe any one hat I wore at SpaceX, because they were switching on and off so fast it didn't seem like there were any hats," Kassouf said.

As Musk drove his employees to work long hours, he created an environment where they would want to be, day and night. The company served lattes and then meals. And each department got a food budget. As the avionics department expanded under Koenigsmann, it moved a block away into 211 Nevada, and received $250 from Musk each week for Costco snack runs. The task rotated among the department's employees. One test engineer named Juan Carlos Lopez made elaborate meals like carne asada. Others opted for the simpler, calorically dense chips and sweets.

Thus, when the avionics crew took a break from working on printed circuit boards, testing hardware, or writing flight software, they could have snacks while they played Ping-Pong.

"You just really needed something to break up the constant push to be doing things a lot faster than what we were doing," Altan said. "If you didn't have that lighthearted outlook at SpaceX, it would have been a really tough time to survive in the early years."

• • •

Musk differed from his competitors in another, important way—failure was an option. At most other aerospace companies, no employee wanted to make a mistake, lest it reflect badly on an annual performance review. Musk, by contrast, urged his team to move fast, build things, and break things. At some government labs and large aerospace firms, an engineer may devote a career to creating stacks of paperwork without ever touching hardware. The engineers designing the Falcon 1 rocket spent much of their time on the factory floor, testing ideas, rather than debating them. Talk less, do more.

There are basically two approaches to building complex systems like rockets: linear and iterative design. The linear method begins with an initial goal, and moves through developing requirements to meet that goal, followed by numerous qualification tests of subsystems before assembling them into the major pieces of the rocket, such as its structures, propulsion, and avionics. With linear design, years are spent engineering a project before development begins. This is because it is difficult, time-consuming, and expensive to modify a design and requirements after beginning to build hardware.

The iterative approach begins with a goal and almost immediately leaps into concept designs, bench tests, and prototypes. The mantra with this approach is build and test early, find failures, and adapt. This is what SpaceX engineers and technicians did on the factory floor in El Segundo, and it allowed them to capture basic flaws with early prototypes, fix their designs, and build successively more "finished" iterations.

An independent company like SpaceX can afford the latter approach, said planetary scientist Phil Metzger. He cofounded the Swamp Works project at NASA's Kennedy Space Center in 2012 to push the space agency toward leaner, more rapid development projects, but ultimately could not break through the government bureaucracy.

"We were always fighting for the recursive, nonlinear approach, which is best early in a program," Metzger said of his NASA experience.

"To adopt this method, you have to let people see you fail, and you have to push back when the critics use your early failures as an excuse to shut you down. This is why it is hard for national space agencies to adopt it. The geopolitics and domestic politics are brutal."

Failure *was* an option at SpaceX, partly because the boss often asked the impossible of his team. In meetings, Musk might ask his engineers to do something that, on the face of it, seemed absurd. When they protested that it was impossible, Musk would respond with a question designed to open their minds to the problem, and potential solutions. He would ask, "What would it take?"

If Musk asked Kassouf to jump a fifty-foot fence, he did not want to hear it was impossible. He wanted Kassouf to ask for a pogo stick with a certain kind of a spring on it, or maybe a jetpack, and get on with it. Musk pushed his engineers to try new approaches to difficult problems. If they had good ideas, he would back them with resources.

After Musk hired a few experienced hands to lead his propulsion, structures, and avionics departments—Thompson, Mueller, and Koenigsmann—he mostly brought on recent college graduates. Most had no significant others pulling on their time, asking when they'd be home for dinner. They lived in apartments, not houses with lawns to mow. They had no children to look after. So they worked long, hard hours as Musk squeezed everything out of them that they had to give. And most were more than willing to give SpaceX the best years of their lives. Musk was a siren, calling brilliant young minds to SpaceX with an irresistible song. He offered an intoxicating brew of vision, charisma, audacious goals, resources, and free lattes and Cokes. When they needed something, he wrote the check. In meetings, he helped solve their most challenging technical problems. When the hour was late, he could often be found right there, beside them, working away. And when they needed a kick in the ass, he deployed his stare, or a few sharp words.

Through it all, Musk kept their focus on launch. Originally, he wanted

SpaceX to launch by the end of 2003. He had schedules posted above the urinals in the men's room. The company failed to make this date, but by the second half of 2003, the glossy factory floor began to fill up with rocket parts. Just two years later, in late December 2005, SpaceX had a rocket on a launchpad, half a world away, counting down to launch. This mad, frantic rush toward orbit began with Musk instilling his workplace culture at the building in El Segundo. He did so by getting his own hands dirty, holding long, technical meetings where ideas flowed freely, and in late-night gaming sessions. Some did not make it. You either fit in and accepted the demanding culture, or you left.

The last thing Musk ever wanted to hear from an employee was "But that's how it's always been done." The members of the growing SpaceX team, both seasoned and green, had all come from somewhere. Those not hired directly from college, especially the technicians, came from big aerospace companies like Boeing or Lockheed, with their own cultures. These big contractors largely subsisted off of government business, with a certain way of doing things to maximize profits while also satisfying the customer's needs. This often involved stretching out contracts, since Uncle Sam was paying for their time. During one early meeting at El Segundo, some former Boeing and Lockheed workers began bantering back and forth about their old companies, and the merits of how things had been done.

Musk raised his voice to end the discussion. "You work at SpaceX now," he sternly reminded them. "You bring that up one more time, and we're going to have serious problems."

The message was clear. Wherever they had come from, whatever they had learned at those places, they were now part of the SpaceX team. Musk had hired them all, personally, to change the world. They had a job to do. A very hard one.

2

MERLIN

August 2002–March 2003

Tom Mueller had high hopes for his forty-second birthday in March 2003. It had been a whirlwind of a year. After joining SpaceX, Mueller dove headlong into designing a new engine for the Falcon 1 rocket. The last six months, especially, had passed in a blur. First, hiring a few key employees. Then moving engine testing halfway across the country, from California to Texas. Finally, his small team had completed the core of the Merlin engine. And now, on his birthday, they planned to light the candle.

Anticipating the very first test of the thrust chamber assembly, the Merlin's heart, Mueller had brought an expensive bottle of spirits to McGregor, Texas. A few weeks earlier, he'd spied a distinctive glass bottle of Rémy Martin cognac on the desk of Mary Beth Brown. It was left over, she said, from a recent space conference. The bottle retailed for $1,200.

Impressed, Mueller asked if he could bring the bottle to Texas for the propulsion team to celebrate the first time they ran the engine. He could. "So I grab this bottle and take it," Mueller said. "This was like a month before the testing, so I put it in the cabinet at McGregor and told the crew, 'Nobody touch this until we run the engine.' So I had it sitting in the cabinet, and we left it alone."

At 9:50 P.M. on March 11, the propulsion team declared the Merlin's thrust chamber, where liquid oxygen and kerosene propellants mix and burn, ready to go. They ignited the engine and ran it for half a second. The Merlin burned as hoped, and shut down without an explosion. This moment well and truly called for a celebration. Mueller passed around small paper Dixie cups, and began pouring Rémy Martin. After a few shots, the team realized they still had to make the twenty-five-minute drive from the McGregor test site back to the apartments they rented on the outskirts of Waco. Back then the small team all rode together, to and from work, in a white H2 Hummer. The test site director, an engineer named Tim Buzza, drew the short straw of driving home for the night along U.S. Route 84.

Most of the trip passed uneventfully, until Buzza slowed to make the turn off the highway, and onto the road leading to their apartment complex. Bright lights flashed in the rearview mirror. A Texas State Police officer turned on the siren of his patrol car, and pulled the Hummer over to the side of the road. It was Texas. They had been drinking. This was bad. Mueller remembers the mood inside the vehicle: "We're like, 'Oh, holy shit. Everybody's going to jail.'"

The officer walked over, grabbed Buzza, and pulled him out of the Hummer. After leading Buzza around to the back of the vehicle, the officer asked accusingly, "All right, what's going on?"

Buzza recalls not being sure what he had done wrong. The Hummer had not been speeding or violating any traffic laws he could think of. Buzza suspected the flashy white Hummer might have drawn the officer's attention; most of the vehicles trundling down the rural Texas highway were beat-up pickup trucks and farm equipment.

For Buzza, the best course of action seemed to be telling the truth. He explained that they had just fired a sixty-thousand-pound rocket engine for the first time, and had worked late in the night to do so. All the people inside the Hummer wanted to do, Buzza told the officer, was drive one mile down the road, go home, and get some sleep.

The state trooper said he had never heard such a wild story before, and that they had better be telling the truth. He sent Buzza on his way, trailing the white Hummer to the apartments, where the propulsion team crashed for the night. It was a birthday Tom Mueller would never forget.

Mueller grew up in St. Maries, Idaho. In this small town about one hundred miles below the Canadian border, nighttime temperatures drop below freezing about half of the year. The high school's mascot is a lumberjack, which offers a clue about the regional economy. Indeed, Mueller's father hauled logs across Idaho for most of Tom's childhood, but by the time he graduated from high school in 1979 his dad had bought a small bulldozer. With this he pulled logged trees out of the forest, a process known as skidding.

As he came of age, Mueller realized he wanted a life other than logging in small-town Idaho. Sure, he'd ridden dirt bikes through the woods with his friends, and that had been fun enough, but he also spent a lot of time in the library, checking out science fiction books. And he had always been interested in rockets. He preferred the Centuri brand to the more commonly known Estes model rockets, shooting them off in his grandfather's field. Despite the reading and the rockets and restlessness, Mueller had no real sense of the world beyond Northern Idaho, nor what he might make of himself.

Then Mueller got a lucky break during high school, in geometry class. The teacher, Gary Hines, noted that his pupil had an aptitude for mathematics, and told Mueller, "Hey, you're really good at math. You're going to be an engineer, right?"

Mueller responded that he thought, maybe, he would become an aircraft mechanic. He liked working with his hands, and he liked things that flew.

The teacher said he thought Mueller could do more: "Do you want to be the guy that fixes the plane, or the guy that designs the plane?" Mueller liked the sound of this. Hines helped ensure that the rest of Mueller's high school class load prepared him for college. So while his friends had work study or easy classes during their senior years, Mueller took calculus and advanced biology.

Another teacher, Sam Cummings, also had a profound influence. Cummings was one of those spirited teachers who touched so many lives, finding the gifts within his students and encouraging their expression. Cummings taught science at St. Maries High School for thirty-four years, winning state and national awards. In particular, he pushed his students to attend regional science fair competitions. Mueller took to this, and built a rocket engine out of his father's welding torch. He ended up qualifying for the International Science and Engineering Fair in 1978, and took his first-ever airplane ride to Anaheim to compete.

After high school, Mueller scraped together enough money to attend the University of Idaho, seventy miles away from St. Maries. But he was not yet quite free of logging. To pay for school he worked summers back home, wielding a chainsaw in the forests. It was backbreaking work, but paid well. He spent long, hot days in the hills, working a strip of trees by first felling them, and then cutting off their tops so they could be hauled out and processed. In summer, the sun rose before 5 A.M. and did not set until 9 P.M. at night. He worked alongside a partner within hailing distance. The rule was that if a chainsaw stopped whirring for a while, you checked on your partner.

One day during the summer of 1982, Mueller felled a tree that landed near a snag, one of the most hazardous activities in logging. Snags are dead trees, tall and white, with no limbs. So as Mueller finished cutting his intended tree, he watched the snag closely to make sure that, with all the jarring, it didn't tumble over in his direction. The snag shook for a moment, but then appeared to steady itself. Mueller exhaled, and bent down to put his chainsaw on the ground for a rest. As he stood up, Mueller glanced back toward the snag. Silently, it had begun falling, and rapidly filled his vision. He leaped out of the way, just in time. "If I hadn't looked up," he said, "I'd be dead." But he had. The tree only grazed his left foot, spraining his ankle.

It took five years for Mueller to get through college, in part because one summer they weren't hiring loggers back home, so he had to take a semester off to raise enough money to finish school. When he graduated, the newly minted mechanical engineer faced a difficult decision on what to do next.

Years before, an uncle had also attended the University of Idaho, and also majored in mechanical engineering. The uncle had gone off to California, taking a job at an industrial building company called Johnson Controls. He had not taken to the work or the location, and soon returned to St. Maries where he eschewed his engineering degree and took employment as the town's garbage collector.

But Mueller was not put off by this. After graduation, a local forklift company extended a job offer, as did Hewlett Packard in Boise. But Mueller did not really want to work for either of those companies. He had launched model rockets since the third or fourth grade. He'd built crude rocket engines. And what Tom Mueller wanted to do more than anything was get out of Idaho and work on real rockets. So he informed his parents that he intended to move to California. In college he had met and married his wife, and her mother lived in Los Angeles. Mueller didn't have a job offer in southern California, but he knew that was where he would find the aerospace industry. His parents thought him crazy. Mueller's father told him there were millions of kids coming out of college. The space shuttle era had just dawned. Everyone wanted these jobs. Why did he think he had a chance of getting one?

"I think that's why Elon liked me, because I was very optimistic,"

Mueller said. "And my dad was really a pessimist, so I don't know where I got the optimism, but I'm just like, 'No, I'm going to go get a job and build rockets. I'm going, I'm doing it. Nothing's stopping me.'"

They said he'd be back. And that the forests would be waiting. But Mueller was never returning. By moving to Los Angeles, he took the first step down a path that would lead him to a job designing rocket engines for a large company, TRW, and then SpaceX. Ultimately, in 2020, the first astronauts to reach orbit from American soil in nearly a decade would do so sitting on top of the rocket engines he designed.

Upon reaching Los Angeles in the summer of 1985, Mueller started mailing out résumés. Not that he had much experience to brag about. He had none, really. And his grades were only so-so. But he sent out dozens of letters and résumés from May through August. Not a single company replied. He tried cold-calling companies he found listed in engineering magazines. Still, he got nowhere. By the end of the summer Mueller was getting desperate. In the back of his mind, he heard echoes of his father's doubts. As the City of Dreams slowly crushed his, Mueller feared he might have to swallow his pride and move back to Idaho.

A breakthrough came that fall, when he attended an aerospace job fair. While his résumé might have lacked distinguishing traits, in person Mueller oozed enthusiasm and knowledge about rockets. He landed three interviews, which turned into three job offers. Rocketdyne offered him a job working on the space shuttle main engines, a prime gig, but the pay was low. General Dynamics invited him to join its Stinger surface-to-air missile team, which intrigued Mueller. It was a good offer, but when Mueller drove out to the company's facility in Pomona, west of Los Angeles, smog engulfed him almost like a forest fire. Not for him. Hughes Aircraft offered him a job working on satellites. It wasn't rocket engines, but the pay was excellent, and he liked the location in El Segundo. He took the job and, after a couple of years, through a friend, Mueller found out about a job opening at TRW. This was a large automotive and aerospace firm that had done a number of interesting things in space, including building the rocket engine that landed Neil Armstrong and Buzz Aldrin on the Moon.

He ended up working at TRW for fifteen years, doing mostly rocket engine development, big and small. In the mid-1990s, Mueller started on the TR-106 project, one of the most powerful engines built in decades. This new, liquid-fueled rocket engine produced 650,000 pounds of thrust, about 30 percent more than a space shuttle main engine. With the TR-106, Mueller hoped to win a contract from Boeing to power the company's new Delta IV rocket. Mueller said TRW estimated it could build engines for about $5 million per unit. Boeing instead selected a different engine costing more than four times as much, built by Rocketdyne. This company boasted a storied propulsion heritage—Rocketdyne was the Nike or Louis Vuitton of rocket engines—and it charged a premium. TRW did not protest. "Our engine ran great," Mueller said. "But the company wasn't behind it. TRW didn't really care about their rockets, they just cared about their spacecraft. They kept the rocket guys there just because they needed them as a necessary evil for their spacecraft."

Around the time he started working on the TR-106 engine, Mueller also began attending meetings of the Reaction Research Society. Founded in 1943, the society had its own test site north of Los Angeles, near Mojave, California. On weekends, club members would drive a couple of hours north of the metro area and fire off their creations. At these meetings Mueller made friends with like-minded enthusiasts, including John Garvey. The pair bonded and eventually codeveloped what may be the world's most powerful amateur rocket engine, which had twelve thousand pounds of thrust. Mueller did most of the technical work, while Garvey provided space in an industrial park, and funding for fuel tanks to feed the engine. They had nicknamed their booster "BFR," short for Big Fucking Rocket.

In January 2002, Garvey told Mueller that an internet millionaire named Elon Musk wanted to come by and meet him that weekend, and see their amateur engine. Mueller didn't think much of it until Musk and his first wife, Justine, walked into the shop a few days later. She was visibly pregnant, but that had not stopped the couple from dressing for an evening on the town. Their elegance provided a sharp contrast to the sweat-stained rocket scientists. As Musk entered, Mueller and Garvey were straining to bolt the eighty-pound engine onto an air frame.

"Have you ever worked on anything bigger?" Musk asked.

Mueller explained about his past work on the massive TR-106 engine, as well as several other propulsion projects. Musk continued firing off questions, drilling into technical details about thrust, injector design, and, most important, cost. What was the least amount of money that Mueller would require to build a powerful engine? Musk wanted to talk more, but the couple had plans. He asked if they could meet again the following weekend.

Mueller hesitated. He had just purchased a Mitsubishi fifty-five-inch widescreen television, one of those bulky sets built on a large cabinet. Back then, HDTVs were the hot new technology, and Fox was producing the Super Bowl in widescreen format for the first time. Mueller and his wife had planned a party, and he wanted to show off the new television to his friends. But Musk got his way, as he, Garvey, and a few other propulsion specialists joined the football festivities at Mueller's house in Long Beach.

"I think I maybe watched one play," Mueller recalls. Three months later he would join Musk at SpaceX.

• • •

n the spring of 2016, Amazon founder Jeff Bezos invited a handful of reporters into his rocket factory in Kent, Washington. No media had been allowed inside before, but Bezos's secretive, fifteen-year-old space company named Blue Origin was finally beginning to reveal the full scope of its plans. Like Musk, Bezos had identified low-cost access to space as the key hurdle standing between humans and moving out into the Solar System. He, too, had begun building reusable rockets.

Over the course of three hours, Bezos led a tour through his glossy factory, at turns showing off Blue Origin's tourist spacecraft, hefty rocket engines, and large 3D printers. He also shared his basic philosophy of *Gradatim ferociter*, Latin for "Step-by-step, ferociously." Rocket development begins with the engine, explained Bezos, who was then working on his fourth-generation engine, known as the BE-4. "It's the long lead item," he said, casually strolling through the factory, wearing a blue-and-white-checkered shirt and designer jeans. "When you look at building a vehicle, the engine development is the pacing item. It takes six or seven years. If you're an optimist you think you can do it in four years, but it still takes you at least six."

In the fall of 2019, as we talked on board his private Gulfstream jet, I related this story to Musk. It was a Saturday afternoon, and we were flying from Los Angeles to Brownsville, Texas. This interview had originally been scheduled for early evening the day before, at SpaceX's factory in Hawthorne, California. An hour past the scheduled time on Friday, his apologetic assistant texted that a crisis had come up. Musk felt terrible, she said, but we would have to do the interview at a later date. I returned to my hotel, preparing to fly back to Houston, when the assistant called back that evening. Musk had decided to visit the company's Starship build site in South Texas that weekend and wanted to know if I cared to tag along. We could do the interview during the flight.

Three of Musk's sons joined their dad for the trip, along with their dog Marvin (as in Marvin the Martian). A well groomed and mannered

Havanese, he adored his master. With Marvin at Musk's feet, we had gathered around a table at the back of the plane, for the interview. Clad in a black "Nuke Mars" T-shirt and black jeans, Musk wanted the boys to hear Dad's stories about the old days.

Musk laughed when told about Jeff Bezos's timeline for engine development. "Bezos is not great at engineering, to be frank," he said. "So the thing is, my ability to tell if someone is a good engineer or not is very good. And then I am very good at optimizing the engineering efficiency of a team. I'm generally supergood at engineering, personally. Most of the design decisions are mine, good or bad." Boastful? Maybe. But SpaceX built and tested its first rocket engine in less than three years with Musk leading the way.

Musk and Bezos, at least, would agree on this much: the process of building a rocket begins with the engine. After all, *engine* is the root word of *engineer*. In principle, a rocket's propulsion system is simple: An oxidizer and a fuel flow from their respective tanks into an injector, which mixes them as they enter the combustion chamber. Inside this chamber, the fuels ignite and burn, producing a superhot exhaust gas. The engine's nozzle channels the flow of this exhaust in the opposite direction a rocket is meant to go. Newton's Third Law of Motion—for every action, there is an equal and opposite reaction—does the rest.

Alas, the reality of building a machine to manage the flow of these fuels, control their combustion, and channel an explosion to lift something toward the heavens is staggeringly complex. And that's not to mention fuel efficiency. A rocket engine's thrust depends on the amount of fuel burning, its exit velocity, and pressure. The greater each of these variables are, the more thrust an engine produces, and the heavier payload it can power into orbit. Conversely, if it takes too much fuel to produce a large enough thrust, or the engine is too heavy, a rocket will never leave the ground.

Musk recognized early on that when it came to propulsion, Mueller was not a good engineer—he was a great one. For the Falcon 1 rocket

Musk wanted a lightweight, efficient engine that produced about seventy thousand pounds of thrust. This, he reasoned, should be enough to get a small satellite into orbit. Mueller had helped design and build several engines at TRW, some more powerful than this, and some less so. The Merlin engine would draw upon some of these concepts and ideas, but Mueller said he and Musk began with a "clean sheet" design. Few of Mueller's friends in the industry believed building a brand-new, liquid-fueled rocket engine without government backing was possible. "All these guys told me a private company can't build a booster engine, that takes the government," Mueller said.

SpaceX did not invent the Merlin engine out of whole cloth. As with almost all rocket engines, the Merlin evolved from previous work. For example, although Mueller had developed a lot of different engines, he lacked experience with turbopumps. Rockets use a staggering amount of fuel, and a turbopump is the machine that feeds propellant into a rocket engine as fast as possible. Inside the Falcon 1 rocket, liquid oxygen and kerosene fuels would flow from their tanks into a rapidly spinning pump, which would spit out this propellant at high pressure, delivering fuel into the combustion chamber primed to produce the maximum amount of thrust. One of the first issues Mueller had to address was how to build a turbopump.

In the late 1990s, NASA had developed a rocket engine nearly as powerful as the proposed Merlin engine called Fastrac. There were other similarities. Fastrac used the same mix of fuels, liquid oxygen and kerosene, a similar injector, and had the potential for reuse. Despite a series of successful test-firings, NASA canceled the program in 2001. Given these commonalities, Mueller thought SpaceX might be able to use the turbopumps NASA built for the Fastrac engine. He and Musk visited NASA's Marshall Space Flight Center in Alabama shortly after Fastrac's demise in 2002, and asked if they could have them. Yes, they were told, but SpaceX would have to go through NASA's procurement program, which could take a year or two. This was too slow for SpaceX, so Musk and Mueller moved on to Barber-Nichols, the contractor that had built the turbopumps.

Barber-Nichols, it turned out, had had a devil of a time building the Fastrac turbopump. To work with the larger Merlin engine, Barber-Nichols would need to do a lot of redesign work. They went back and forth with the SpaceX engineers. During one visit to the Colorado-based company, a designer there happened to suggest a name for the engine to Mueller. Musk had chosen the Falcon name for the rocket, but said Mueller could name the engine, stipulating only that it shouldn't be something like FR-15. It should have a real name. One Barber-Nichols employee, who was also a falconer, said Mueller should name the engine after a falcon. Then, she began listing various species of the bird. Mueller chose the merlin, a medium-sized falcon, for the first-stage engine. He named the second-stage engine after the smallest of falcons, the kestrel.

When Barber-Nichols finally delivered the redesigned turbopump to SpaceX in 2003, it still had major problems. This forced Mueller and his small team to begin a crash course in turbopump technology. "The bad news is that we had to change everything," Mueller said. "The good news is that I learned everything that can go wrong with turbopumps, and really how to fix them." Because the pressurization of rocket fuel allows an engine to squeeze out a maximum amount of thrust, good turbopumps are essential. This would become one secret to SpaceX's eventual dominance of the global launch market. Mueller said the original pump from Barber-Nichols weighed 150 pounds, with an output of about 3,000 horsepower. Over the next fifteen years, SpaceX engineers continued to iterate, changing the design and upgrading its parts. The turbopump in the modern-day Falcon 9 rocket's Merlin engine still weighs 150 pounds, but produces 12,000 horsepower.

. . .

Mueller's original team at SpaceX was small. While he worked through the technical details of the Merlin engine, the company needed someone to find and build a test site. It had to be remote because when it comes to rocket engines, bad things often happen. "Rocket engine development is really ugly early on," Mueller said. "It always is. There's always so many things that can go wrong, and when they do, it's usually pretty catastrophic."

Mueller helped Musk hire Tim Buzza in August 2002, to establish a test site where the company could safely blow things up. Buzza had grown up in Pennsylvania steel country during the late 1970s, the grim, real-world setting for the movie *Deer Hunter.* His father owned a machine shop, where Buzza and his brothers would work for four hours each day after school. Buzza learned to program machining tools in ninth grade, and showed enough aptitude to go on to college at Penn State. There, he fell in love with rocket engines. Over the course of a fourteen-year career at McDonnell Douglas and Boeing, Buzza specialized in pushing the limits of airplane and rocket components to identify potential failures. He'd worked on the large military C-17 transport aircraft as well as the Delta IV rocket. When his former Boeing colleague Chris Thompson called Buzza about working at SpaceX in the summer of 2002, he had just remodeled his home, doubling his mortgage payment, and his wife had shut down her business with the birth of their second child. "It's crazy to think how I rationalized leaving a corporate job at Boeing for a start-up with a guy I never met," Buzza said of Musk.

Buzza set about establishing a test site where the company could put the Merlin engine through its paces. The Mojave Air and Space Port north of Los Angeles offered facilities for engine testing. No place in the world is really like it, combining an airport, a spaceport, an airplane graveyard, and rocket development warehouses. It was where, in 2004, Scaled Composites' SpaceShipOne made the first privately funded human spaceflight. In the 2000s, other companies such as XCOR, Masten

Space Systems, and Virgin Galactic called the facility home for their development operations. It seemed an ideal location for SpaceX, and initially Buzza reached a deal to use some of XCOR's property and equipment for tests.

The space was soon needed. By the fall of 2002, Mueller had already built a prototype for the Merlin engine's gas generator. This is, itself, a small rocket in which oxidizer and excess fuel burn to provide hot, sooty exhaust to drive a turbine. In turn, this powers the turbopump and supplies energy to the heart of a rocket engine. If you want to bring a rocket engine to life, the process starts with the gas generator.

To assist with this, Buzza recruited another engineer from Boeing, Jeremy Hollman. He was younger, only twenty-four, and a Midwesterner who had just graduated from Iowa State University in 2000. But in his time at Boeing, Hollman had earned Buzza's trust as they worked side by side. After joining SpaceX in September, Hollman grew into the role of Mueller's chief propulsion lieutenant. When they started testing the gas generator in October, Hollman would write procedures for the day's testing activity and work on the hardware. Buzza saw to the ground support equipment and wrote software. And Mueller directed the operations, with overall responsibility. Hollman described working under Mueller as "a doctoral-level education in liquid propulsion."

Some of the early tests with the gas generator were pretty hairy. As fall turned to winter at Mojave in 2002, Hollman said they performed the first, full ninety-second test-firing. This produced a huge, black, and smoky cloud over the spaceport. Almost exactly as the test ended, the wind died. "The cloud parked itself literally right around the flight control tower," Hollman said. "You could not have parked a sooty, smoky cloud in a worse spot at the airport."

As Mueller churned out designs for a Falcon 1 gas generator and other propulsion elements, someone had to build them. Mueller had contracted with a local machining firm, Mustang Engineering, while at TRW. Now, at SpaceX, he started to email them drawings and PDFs. "Man, those guys would send me some of the wildest crap I'd ever seen," said Bob Reagan, co-owner of the company. SpaceX also paid quickly. Within a day of receiving a purchase order from SpaceX, Reagan would have a check. Initially, Reagan tried to explain to Mary Beth Brown how things typically worked. With other companies, Reagan told Musk's assistant, he would finish a part, submit an invoice, and receive a check thirty days later. She was unfazed. SpaceX wanted its parts fast. Reagan got the message and began to prioritize Mueller's orders.

Reagan had started machining right out of high school, in 1982, in southern California. During his early career he'd fabricated components for the space shuttle, working on its solid rocket boosters, and helped make the cradle that held the Hubble Space Telescope inside the orbiter's payload bay. Reagan had also taken a lot of orders from Boeing as it manufactured the Delta IV rocket, and other large aerospace companies.

Reagan had never worked with a company that moved as quickly as SpaceX. He'd receive orders from them and, within a few days, ship parts machined out of aluminum or other materials. But one day in the fall of 2003, when Hollman called and said he needed a particular part rushed over, Reagan replied that he could not help. He and his business partner had fallen out, and the only course was to shut Mustang Engineering down. SpaceX would have to get its parts from elsewhere. Considering SpaceX's other options for machining services, Hollman despaired. SpaceX had come to rely on Reagan's promptness. Hollman urged Reagan to come and meet with Musk.

Mueller was not sure that would be a good idea. He did not know how Musk would react to Reagan, who spoke gruffly, had long hair, wore earrings, and rode Harleys. But Mueller and Hollman eventually agreed the machinist should come in for an interview. They need not have worried. Reagan could clean up when he needed to. He'd done work for Boeing, after all, and if you were going to meet with Boeing you had better wear a tie.

Not that Musk cared about ties. Or earrings. He just wanted someone who could build things, fast and cheap. During the interview, Musk explained that he was paying for everything at SpaceX out of his pocket. And then he asked Reagan, "What's the cheapest you would work for?" They haggled for a time, but eventually Musk agreed to meet Reagan's asking price. He needed the help. Ten minutes later he returned with a contract. It was Saturday, November 1, at 5 P.M. Musk wanted his new vice president of machining to start work that evening.

Soon enough, Reagan had his shop at SpaceX humming. He hired six of Mustang's machinists, and Musk bought the machines his company was about to liquidate. By bringing Reagan in house, Musk essentially cut much of his manufacturing costs in half. Now he could buy a chunk of aluminum and have people in his building work it as if it were clay, producing a part on demand without the markup and delay of sending it to an outside machining shop. And the lines of communication between SpaceX's engineers and the manufacturing crew were wide open.

"Before, if I had a problem with one of my customers, I'd have to call the buyer," Reagan said. "Then the buyer would call the engineer, and a week later I might get a call back with an answer." At SpaceX Reagan sat in the cube farm. If the engineers did something he thought was dumb, or would not work, he would tell them. There was mutual respect.

His relationship with Musk was simple. "He can't stand a liar, and he hates a thief," Reagan said. "And if you say you can do something, you'd better fucking do it." Apparently Musk liked what he saw from his new machinist, who had begun to put in seventy-five- or eighty-hour weeks at the factory in El Segundo. Less than a month after Musk hired Reagan, he took him to lunch in his McLaren F1 sports car. Before they had left the parking lot, Musk told Reagan that he was getting a $10,000 raise.

"These guys said you were good," Musk told him. "I didn't know you were *that* good."

Reagan had a front-row seat for a dispute between Mueller and his former employer, TRW. Shortly after Mueller left in 2002, aerospace giant Northrop Grumman acquired the company's missile development and spacecraft divisions. Northrop is one of the world's largest weapons manufacturers, but also does civil aerospace work. Boeing and Lockheed Martin are NASA's two biggest legacy contractors for space work, but Northrop ranks close behind.

The issue between Mueller and his former employer concerned the Merlin engine's use of a special design for its injector. This part of the engine controls the flow of oxidizer and rocket fuel into the combustion chamber. If there's too much fuel, then wasted, unburned propellant leaves the engine. Too little, and the thrust is lower. There are several different types of injectors to regulate propellant flow and mix them just before ignition. Mueller began from a design he was familiar with, called a pintle injector, which is somewhat like a coaxial cable. For the Merlin engine, the liquid oxygen flowed through the middle, fanning out as it emerged like the spokes of a wagon wheel. The kerosene fuel moved through the outside of the pintle, and came out as a sheet. "We called it the sheet hits the fan," Mueller said.

The problem, from Northrop Grumman's perspective, was that TRW had invented the pintle injector for the Lunar Module's descent engine in the 1960s. After learning of the Merlin engine's modified design, Northrop sued in California state court, alleging Mueller and SpaceX had stolen trade secrets. SpaceX countersued, saying Northrop had used its position in an advisory role to the Air Force to essentially spy on SpaceX. Reagan said the Northrop lawsuit was ridiculous because SpaceX must have gone through fifty different designs for the Merlin's injector, iterating all the while, blowing things up, and keeping his machinists busy in 2003 and 2004. "It seemed like a big joke to me," he said of the lawsuit.

By early 2005 the companies agreed to drop their mutual lawsuits. Neither admitted wrongdoing, or paid legal fees or damages.

SpaceX likes to operate on its own terms and its own timeline. It wants to experiment, iterate, and, on occasion, make huge clouds of black smoke that may errantly stray over flight control towers. Anything that slowed down these explosive efforts represented a hindrance to plow through. Late in the fall of 2002, therefore, it became clear that working in the relatively controlled environment of Mojave Air and Space Port would not suit.

As the propulsion team tweaked the Falcon 1 rocket's gas generator, Buzza deepened his discussions with the spaceport general manager, Stu Witt, for a long-term lease at Mojave. Witt already had expressed environmental concerns about what SpaceX wanted to do. But now, during the final phase of negotiations, he sought to limit the scope of their testing to an engine with thirty thousand pounds of thrust. Smaller engines produce smaller explosions. But this was less than half the thrust SpaceX expected the Merlin engine to ultimately produce.

"Here I was, standing on a plot of land with stakes in the ground, and a set of plans in my hand going, 'This is just a nonstarter,'" Buzza said.

SpaceX would not be going to Mojave after all. That fall, Musk and his propulsion team considered a handful of other sites, including an area of Edwards Air Force Base in California near Mojave, as well as traditional test sites at NASA centers in Alabama and Mississippi. But really, they preferred a site unfettered by government control. They all knew of a rocket company called Beal Aerospace that had gone bust a couple of years earlier. Photos of the company's now-defunct test site in McGregor, Texas, were still online. It looked promising.

In November 2002, Purdue University invited Musk to speak. It was a prestigious invitation, as Purdue's notable aerospace engineering program has churned out dozens of NASA's most notable flight directors and astronauts, including Neil Armstrong. Musk brought some of his senior team to recruit graduating students. While there, Buzza and Mueller chatted up a professor named Scott Meyer, who had been a senior propulsion engineer at Beal. To answer more of their questions, Meyer shared the name and number of Joe Allen, a former employee at the McGregor test site who still lived in the area.

Musk decided, that day, to fly down to Texas to check out the site. En route, Mueller placed a satellite telephone call to Allen, to see if he would meet them there and give a tour. Allen had been Beal's last employee, staying on for eight months after the company's end in October 2000, to look after the site while assets were liquidated. A native of nearby Meridian, Texas, and with a background in machining, Allen had been working in the trade for three decades by then. But during his three years with Beal, Allen had seen how everything with rockets ran on computers, so after getting laid off he had begun to take programming classes at Texas State Technical College. He was in the midst of an exam when Mueller called, but agreed to meet them. Look for me under the big tripod, Allen said. Can't miss it.

Driving across the flat expanses of Central Texas to the rocket testing site near the small town of McGregor, Musk and the others could not help but be impressed by its remoteness. The towering concrete tripod was easy to find, and true to his word Allen awaited them beneath it, standing outside his old blue Chevy pickup truck. The site suited SpaceX's needs because Beal had developed it specifically to test rocket engines. As Allen toured Musk and the others around, he showed them not only the tripod test stand, but also a large bunker for monitoring engine firings and other facilities. The town of McGregor owned everything here and was willing to lease it all. Because local officials wanted the company as a tenant, there would be minimal interference, and no restrictions on the size of their engine. Texas, too, had a much less restrictive regulatory environment than California, with more business-friendly laws. Musk hired Allen on the spot, and leased the site. "Really, it was perfect for what we needed," Buzza said. "It had everything we could have hoped for."

It lacked only proximity. McGregor is 1,400 miles from Los Angeles. The closest commercial airports of any size were a two-hour drive away, in Austin and Dallas. It would take the propulsion team a full day to travel there, and another day back. No problem, Musk told Mueller. They could have the use of his smaller Cessna jet, and fly right into the tiny McGregor airport with its single runway. Eventually, when the small propulsion team landed, Allen would come to be waiting for them in the white Hummer. "We called it the cattle haul to Texas," Mueller said.

As 2002 faded into 2003, his company just half a year old, Musk found himself with room to roam in Texas. There, his engine designers would construct a new test site. One with wide open spaces and few rules. Over the next two years, Mueller, Buzza, Hollman, and a handful of others would hammer together the Merlin engine, and put it through its paces. They would burn up thrust chambers, blow apart fuel tanks, and raise enough ruckus to bring the Secret Service to their gates. But by 2005 they would build something powerful, almost from scratch, with enough thrust to send half a ton screaming toward outer space.

And this is just what the Merlin engine did on the first flight of the Falcon 1 rocket. For thirty-four seconds, at least.

3

K W A J

January 2003–May 2005

Their day began early, but the SpaceX launch team had no difficulty scrambling out of their beds on the morning of May 21, 2005. As they dressed at a Holiday Inn Express in Lompoc, the dozen or so engineers and technicians were not sure what that afternoon held, but it felt full of promise.

After a previous attempt had ended without success, these employees hoped a second try would bring the Falcon 1 rocket shaking into life for the first time, rattling the California coast north of Los Angeles. It had been three years since SpaceX's founding, and through long days and sleepless nights they had managed to design and assemble a rocket. The company had also navigated the Byzantine Air Force rules and requirements to obtain permission to ignite the booster for a critical test.

That morning, one team of engineers broke off toward a small command center a few miles from the launch site. They had converted a forty-foot truck trailer into a launch control center with a dozen workstations, and called it the Command Van. The larger group drove down to the pad near the ocean to prepare the rocket for its fiery debut. During a static fire test, powerful clamps hold down the rocket while its engine ignites and burns for several seconds, as if it were lifting off from the pad. Such a test ensures a rocket's readiness before an actual launch.

SpaceX had been fortunate to find a launch site so close to its factory, only about a three-hour drive north of Los Angeles. More than half a century ago, the Air Force began using this mountainous terrain along the coast to launch missiles over the Pacific Ocean. Later it grew into a major U.S. spaceport. Vandenberg Air Force Base, named after a general who had helped plan the D-Day invasion, superintended hundreds of launches by Boeing's Delta boosters, Lockheed Martin's Titan and Atlas rockets, and those of a few other firms. Now the brash newcomer, SpaceX, had earned a seat at the table with the big boys. Or so it thought.

As the company's engineers and technicians worked at the launch site, their brilliant white-and-black rocket rose above them. Gradually, in ones and twos, they evacuated from the pad, so that liquid oxygen and kerosene could be pumped into the rocket's fuel tanks. Soon, packed inside the Command Van, wearing headsets, they watched video feeds of their rocket as it puffed clouds of frosty oxygen into the atmosphere. And they got so close. Invariably that day, the countdown clock would tick down to the final seconds, and the onboard computer would sense something not quite right, perhaps the pressure in the engine was too high, or the temperature too hot. Automatically, the computer would trigger a shutdown of the Merlin engine before it ignited.

After multiple attempts, they ran out of liquid oxygen to fuel the rocket. Frustratingly, the truck carrying their final load of this chilled oxidizer had gotten turned around coming out of Los Angeles. For want of oxygen, they had to scrub for the day. As night began to fall over the Pacific coast, the SpaceX launch team that had worked so hard to reach this moment felt heartbroken.

"It was the first time, really, that we weren't completely successful achieving something, and we were embarrassed," said Anne Chinnery, an engineer who had helped spearhead SpaceX's construction of the launchpad at Vandenberg. "We all took it kind of hard."

After they shut down the rocket, and locked up their facilities, the tired and dejected team drove down out of the hills and into the nearest town, Lompoc, to drown their sorrows. To help numb their pain, an engineer named Josh Jung bought shots of Goldschläger. "We all showed up at a bar and boy, we got toasted," Chinnery said. "I'll never forget it. That is probably the most drunk I've ever been in my whole life." For a time, the SpaceX launch team even lost Chinnery. But eventually the group found her, and all returned safely to their hotel, exhausted and dejected, to collapse back into their beds.

The next morning, an unhappy Musk called Tim Buzza, the Falcon 1 test and launch director. Hungover, and bleary-eyed from the events of the day before, Buzza listened as Musk pushed for another attempt the next day. It took some doing, but Buzza eventually convinced his boss that the company's employees needed a little time to recuperate. They were all exhausted after days of effort.

After the phone call, Buzza told his launch team to go home to Los Angeles. They had two days to rest before they would return for another go at conducting the test. That afternoon, he took his own advice and drove back along U.S. Route 101. He stopped at the iconic Santa Claus Lane, a few miles east of Santa Barbara, to take a break.

"The sun was setting, so I went out to the beach," Buzza said. "I lay down and fell asleep for hours, and woke up in the dark and salty chill."

When Neil Armstrong took humanity's first step on another world in 1969, Anne Chinnery's parents woke the sleeping three-year-old to experience the moment. The only conscious memory she retains of the Apollo 11 Moon landing is confusion, but it might nonetheless have proven formative. Throughout her childhood, she maintained an interest in space, and aspired to fly there one day. It became more than a theoretical idea in the early 1980s; as Chinnery neared high school graduation, NASA sent Sally Ride into space.

By then Chinnery's family had moved to Colorado, and she opted to attend the nearby Air Force Academy both out of a sense of adventure and for its potential as a pathway into space. For more than a decade after she earned a degree in astronautical engineering, the Air Force became her home, affording Chinnery some of that adventure she craved. She helped build satellites, assessed the threat posed by foreign-made ballistic missiles, and spent time at Vandenberg Air Force Base assisting launch companies.

But by the turn of the century she had grown restless. In her early thirties, Chinnery could stay in the Air Force for another decade and earn a healthy pension, or she could look for a job in the private sector with less paperwork and more engineering.

On a whim, she attended a conference on lunar colonies and exploration, where she met James Wertz, then president of a small aerospace company in Southern California named Microcosm. In 1999 Wertz hired her, and soon Chinnery befriended coworkers including Gwynne Shotwell and Hans Koenigsmann. Then, her friends began leaving. Koenigsmann departed first in May 2002, followed by Shotwell a few months later. Chinnery decided to leave Microcosm in September of that year, too. Feeling burned out, she felt like she needed a break from the rocket business.

Her friends now working at SpaceX kept reaching out to her, however, and later that fall they persuaded Chinnery to come to the El Segundo office for an interview. Chinnery, they knew, had a background that could help the fledgling company when it came to finding a launch site for the Falcon 1 rocket. Chinnery's Air Force experience could smooth the way toward gaining access and necessary military approvals for Vandenberg. Musk, the great decider of new hires, was not sold. He came away from his interview with Chinnery unimpressed. "I've never met a man so laser focused on his vision for what he wanted," she said of the experience. "He's very intense, and he's intimidating as hell. It's tough when you're interviewing with him." No offer was forthcoming.

But her friends didn't give up, and in early 2003 Chinnery joined SpaceX as a consultant. Soon, the time she had spent at Vandenberg as an officer a decade earlier was paying dividends for SpaceX. She knew people there and how the base operated. Chinnery hired on full-time later that year. For once, Musk's instincts had led him astray.

"Early on, Elon didn't really understand the importance of inter-facing with outside agencies, or how difficult it would be," Chinnery explained. "There was this whole oversight the Air Force wanted over the design, development, and launch aspects of rockets. He really had no idea about that, but it was my specialty."

SpaceX sought to fly missions from Vandenberg because it is by far the closest launch site to the company's factory, and rockets lifting off from there can travel very nearly due south without overflying land. This is perfect for putting satellites into polar orbits, in which a spacecraft flies over the south pole, and then the north pole. As the Earth rotates below, a satellite in a polar orbit can observe the entire planet over the course of a day. With the Falcon 1 rocket, SpaceX aspired to launch small, com-mercial satellites, many of which fly polar orbits for this reason.

After joining SpaceX, Chinnery helped compile paperwork needed to gain access to Vandenberg, and secure the company a launchpad on the sprawling base. Gradually, the former Air Force officer also melded into the company's Silicon Valley work ethos of long days and nights. She bonded with the other cubicle denizens in El Segundo by playing *Quake*, *Doom*, and other games in the evenings. The male-dominated workplace didn't bother her. "At the time, aerospace was predominantly male, and I was very used to being in environments where I was the only female," she said.

In early 2004, the Air Force agreed to let SpaceX use part of Space Launch Complex 3 at Vandenberg, more commonly known as "Slick-3," for SLC-3. The site consisted of two pads built in the late 1950s for some of the first Atlas rocket flights. By the early 2000s, boosters only infrequently launched from the "East" pad, while the smaller "West" site had fallen entirely into disuse. SpaceX could use the western pad, where the launch facilities were mostly demolished. Only a small concrete building and a flame duct, which channels heat and exhaust away from the rocket, remained at SLC-3 West.

At the time, U.S. rocket companies followed a stately, unhurried launch process, often rolling a booster out to the pad months before liftoff. For a company seeking to turn space launch into a commodity, this would not do. SpaceX aspired to build rockets in its factory, and immediately ship them up U.S. Route 101 to Vandenberg on modified semi-trailer flatbeds. Once on-site, in a matter of hours or days, SpaceX planned to move its rocket out to the pad and, like a clock hand moving from nine to noon, raise the booster to point skyward. Then the launch director would light the candle and go. That was the vision, at least.

The final decision, to go or not, would rest with Buzza. As he well knew from his previous experiences with new rockets, SpaceX would not be able to launch on demand from the get-go. Initially, each booster would require weeks of inspection and preparation before liftoff. This proved especially problematic at Vandenberg because the exposed launch site lay less than a mile from the Pacific Ocean. To protect the rocket, SpaceX bought a large, framed building covered with a durable fabric. Then they realized they'd need the tent building to roll away from the rocket before launch, so that it would not be incinerated. Because the tent was not designed to roll, the inventive engineers added wheels. Still, it did not roll easily. So the launch team built rails. And now they had a movable building to protect the Falcon 1.

Until a storm hit. Two days after Christmas, in 2004, winds whistled in off the Pacific Ocean, gusting above 50 m.p.h. across the spaceport. Buzza was at home in Seal Beach, enjoying the holidays and a brief reprieve from work with his family, when he got a call from the company's manager at Vandenberg. An Air Force official had found their tent blown down the hill. It had gone off the rails. Literally.

"I hardly get a break, and now it's Christmas, and I've got to take off and drive up there," Buzza said. He and the site manager, Chip Bassett, spent the rest of the day moving the tent back into place. "We had to pick up the tent with these two SkyTrak lifts and put it back up on the rails, and then secure it more properly. That was just an example of how we were running a little bit fly-by-night with some of our stuff. It wasn't as robust as it needed to be."

As 2004 turned into 2005, Buzza, Chinnery, and the rest of the launch team did plenty of other work at Vandenberg, installing electrical lines across the site, linking their command and control equipment to the launchpad, pouring concrete so that tanker trucks loaded with liquid oxygen could get onto the site, and more. There were a hundred other odd things that had to be installed, assembled, or otherwise prepared for the rocket. And then they were ready. In the spring of 2005, the company trucked its first complete rocket to Vandenberg.

From the beginning, Musk understood SpaceX could not become a sustainable, profitable business from government launch contracts alone. Although the prospect of a low-cost, launch-on-demand rocket appealed to the U.S. military, it had only so many spy and communications satellites ready to fly. To make money SpaceX needed to broaden its customer base to so-called "commercial" customers. This included private companies that wanted to fly satellites to image the Earth or for other business purposes, as well as nations that did not have launch industries of their own.

SpaceX picked up its first real commercial customer when officials with the Malaysian government approached the company in early 2003 to inquire about the Falcon 1. They wanted to know if it could lift a four-hundred-pound Earth observation satellite they were building. The country launched its first microsatellite, TiungSAT-1, on a Russian rocket in 2000. Now, it wanted to put a larger satellite named RazakSAT into a near-equatorial orbit. Malaysia lies just a few degrees north of the equator, and this would allow the satellite to overfly the country more than a dozen times a day.

The mission posed several problems for SpaceX. From Vandenberg the Falcon 1 could not launch satellites eastward, into an equatorial orbit, because rockets aren't permitted to fly over U.S. land. So the company needed an eastward-facing launchpad. Moreover, RazakSAT was too heavy for the early version of the Falcon 1. To put that much mass into orbit, the rocket would need to launch very near the equator and piggyback on the planet's rotation. Satellites lifting off due east from the equator begin with a 1,000-m.p.h. head start on the way to orbit. Effectively, this means a rocket launching from a low latitude can lift more mass than the same rocket from a higher latitude. From a conventional launch site like Kennedy Space Center, at 28.5 degrees north latitude, the Falcon 1 didn't have enough oomph to get RazakSAT all the way into a stable orbit.

"This was a launch deal they were ready to sign," said Gwynne Shotwell, then the vice president for sales, who was eager to reach an agreement. "They had six million bucks. We wanted to sign them, but we had to find a launch site near the equator."

Shotwell's cubicle was near Koenigsmann's, and together in the spring of 2003 they looked at a Mercator projection map of the world. Koenigsmann traced his hand west along the equator from the coast of California. It was all ocean until he got to the Marshall Islands, some five thousand miles away. As they looked at the sprawling chain of tiny islands, Shotwell recognized the Kwajalein Atoll. She remembered something about fighting there during World War II, and felt pretty sure the U.S. military maintained a presence.

In fact, the Kwajalein Atoll had briefly been a focal point in the Pacific theater, when eighty-five thousand men in the U.S. Army and Marines landed on Kwajalein, Roi, and Namur islands in early 1944. After hard fighting, the U.S. forces captured their first foothold in the scattered Marshall Islands, opening the way to further assaults on larger targets like Guam. Following the war, the U.S. military used the islands to stage nuclear weapons tests, and in 1964 the Army established a base there. Later, the military built a missile range known as the Ronald Reagan Ballistic Missile Defense Test Site. The entire facility fell under the jurisdiction of the U.S. Army Space and Missile Defense Command, located in Huntsville, Alabama. There, a lieutenant colonel named Tim Mango had responsibility for Kwajalein.

This tickled Musk. "What are the odds?" he asked. "I sometimes wonder if it's like *Catch-22* where there's somebody doing assignments for majors and colonels, and they said you know what would be funny? If we took Lieutenant Colonel Mango and put him in charge of a tropical island." Musk picked up the phone and called Mango at his desk in Alabama.

To Mango, the call came out of nowhere. After the caller identified himself as someone named Elon Musk, he proceeded to explain in his slightly foreign accent that he was a millionaire who had sold his interest in PayPal and gotten into the space business.

"I listened to his pitch for two minutes, and then hung up on him," Mango said. "I thought he was nuts."

After the call, Mango decided to Google Musk. He found a photo of him with his hand resting on his McLaren F1 sports car. The news article mentioned that Musk had founded a new space company called SpaceX and included a link to the rocket firm's home page. Mango clicked through, and read a little bit more about the company. Maybe this Musk fellow had been serious after all? On the SpaceX website, Mango found contact information for SpaceX, and called the number listed. Someone answered almost immediately in that same, distinct voice. After Mango reintroduced himself, Musk asked, "Hey, did you just hang up on me?"

Mango explained that he had business soon in Los Angeles, and he agreed to visit the SpaceX offices in El Segundo. When he came by the mostly empty facility, he was surprised to find Musk sitting out in the middle of the office, amid a dozen or so employees. They talked for a while, and then Musk invited Mango to dinner at a high-end Los Angeles restaurant. The meal would be well above the Army officer's per diem, however. For the first time in his career, Mango had to call an Army lawyer with an ethics question. The Army officer was told that if he went to dinner, he would have to buy the entire, expensive meal. "I think we went to Applebee's instead," Mango said.

A month or so later, discussions continued during meetings at the Army's Redstone Arsenal in northern Alabama. Musk and a few other SpaceXers flew into Huntsville, and Mango reciprocated the dinner invitation. Huntsville could not match the sophistication of Southern California's restaurant scene, but it did have authentic local flavor. The Army officials decided to take the SpaceX team to the Greenbrier Restaurant, which had no frills but excellent Southern cooking. They urged Musk to try the catfish, and he obliged. Musk was promptly served a whole fried catfish, with the head intact. Not entirely amused—certainly not as amused as the locals in the homey restaurant—Musk ate the catfish.

In June 2003, Musk dispatched Chris Thompson, Koenigsmann, and Chinnery to Kwajalein to assess its potential as a launch site. Accompanied by Mango, they flew 2,500 miles from Los Angeles to Honolulu, and spent the night at the Hilton Hawaiian Village hotel. From Gate 14 at the Honolulu airport, Continental Airlines operated a three-times-a-week flight that made several stops in the Marshall Islands, including

Kwajalein. The airplane took off at 9:30 A.M., and the rocket scientists flew another 2,500 miles to the largest and southernmost island in the atoll, also named Kwajalein.

From the air, the islands were stunning, a string of tiny pearls amid a turquoise sea. Ninety islets comprise Kwajalein Atoll, but their combined landmass is just six square miles, or about one-quarter the size of Manhattan. Each of the coral-covered islands rises only slightly above sea level, forming a broken chain around the world's largest lagoon.

The Army welcomed the visitors with open arms. At the time, the armed forces generally funded about 60 percent of the budget for its major ranges, and expected officers administering the facilities to account for the other 40 percent of needed revenue through commercial contracts. The salty, tropical environment of Kwaj ravaged Army infrastructure on the islands. So Mango and other Army officers were constantly scrounging for external users of the range who could pay fees for radar, telemetry, and other support.

"During that trip they really put out a nice spread for us," Chinnery said. "They treated us like VIPs." Although there were really no dining options on the island other than a military cafeteria, the Army officers did their best. They set a table with fine china and tablecloths, and brought out special food. A photo shows the SpaceX trio standing on the beach, smiling widely, and a bit sunburned. As part of the visit, the Army officials also took the SpaceXers on a helicopter tour, by far the best way to survey the 270-mile atoll.

Later, Musk would also visit Kwaj to assess its potential, and took the same island tour in a helicopter. "It felt like *Apocalypse Now*," Musk said, noting the iconic movie scene in which Robert Duvall's character leads an attack with a squadron of helicopters. "We literally flew in there on a Vietnam-era Huey. We had the doors open. All we needed to do was play 'Ride of the Valkyries.' That was the only thing missing. I'm like, do you have a sound system on this thing?"

As SpaceX employees sized up potential islands to launch from, a tiny speck of land about twenty miles north of Kwaj appeared to offer the best option. Although it measured only eight acres, equivalent to about two New York City blocks, the island was just large enough. It also offered a perfect location, as nothing lay to the east but open ocean for thousands of miles. And perhaps most important, it neighbored the larger Meck Island, which the Army used as a missile test site. A large catamaran took civilians and military personnel between Kwajalein and Meck on a daily basis, and the boat could also stop at Omelek for SpaceX employees.

Situated in the middle of the Pacific Ocean, Chinnery had come about as far on Earth as one could get from a major landmass. At the time, she did not feel overwhelmed by the prospect of building a launch site in one of the most remote places on the planet. "I had kind of caught that weird SpaceX bug that says anything is possible," she recalled. "I was actually thinking how cool it would be to launch from there. Kwaj really is gorgeous. I've never seen prettier water anywhere. I've never snorkeled anywhere better. It didn't really occur to me what a challenge it would be to bring in everything that we needed."

And really, it all seemed like such a distant prospect in 2003, standing on a rocky beach half a world away from home. Someday, sure, the company might build a second launch site in this otherworldly paradise. It might even blast the Malaysian satellite into its equatorial orbit. But not anytime soon. For the Falcon 1 rocket, the pathway to space led into the hills north of Los Angeles, not the coral reefs of Kwajalein. And so the SpaceXers flew home to California, trading tides and waves for traffic and work.

SpaceX attempted its first static fire test at Vandenberg during the first week of May. They found software bugs and bad instrumentation. Their second attempt followed a couple of weeks later, when they drowned their sorrows in Goldschläger. The problem was liquid oxygen. They kept running out of it.

The Air Force, understandably, has strict rules that prevent people from approaching a rocket if both its fuel and oxidizer tanks are full. Fully fueled, a rocket is essentially a bomb waiting to go off. So after the rocket's engine aborted during a test, Chinnery and the other engineers could not simply jump into a pickup truck and go roaring down to the pad to inspect the engine computer's firmware. Instead, they would have to offload the liquid oxygen by dumping it into a gravel bed nearby.

Oxygen condenses into a liquid at a very cold temperature, -297 degrees Fahrenheit, only marginally warmer than the surface temperature of Pluto. This provides challenges in handling and storing cold, or cryogenic, fuels. If you've ever watched the countdown of a rocket launch, the white gas venting away from a rocket is usually liquid oxygen boiling away from the fuel tank. But the payoff is worth it. Because gases take up less space in liquid form, using cryogenic oxygen allows for rockets to have smaller and lighter fuel tanks. And liquid oxygen is a potent oxidizer, combining with rocket fuel to combust rapidly and burn vigorously.

Only after they had dumped the liquid oxygen, or LOX as it is commonly called, could the SpaceXers approach the rocket. Then, having addressed the problem that had tripped the flight computer, they would begin anew the cumbersome process of transferring liquid oxygen back into the fuel tank. At the time, the company hired mobile tanker trucks to bring liquid oxygen to the rocket, and because of boil off during the transfer process, each truck was good for only one or two refuelings. When Musk called Buzza on that morning in late May, he was furious about the LOX. He said if the launch team ever ran out of liquid oxygen again, they would all be fired. "The running joke after that was that we always had to have two shit tons of LOX," Chinnery said. From that point on, a line of at least two or three mobile tanker trucks would be backed up at the launch site, ready to be called upon if needed.

The early difficulties are understandable. With any new rocket, no matter how well designed and engineered, problems invariably arise when individual components are combined into the whole vehicle. It takes time to identify all of these kinks and work through the issues.

Finally, on May 27, everything came together. A thick fog shrouded the rocket that morning, but the countdown proceeded smoothly. When the clock hit zero, the rocket began to rumble. The ambient fog and smoke from the engine shrouded some of the action, but there could be no mistaking what happened on the launchpad that day. For the first time, the Falcon 1 rocket burst into life. It burned bright and hot and loud.

There was but a single casualty. For months, SpaceX employees had chased away a small barn owl that lived in the large duct built to divert flame away from the rocket engine as fuel burned. As the Merlin engine lit, the owl came flying out of the flame trench, but could not avoid being badly singed. The hardy owl had remained in its nesting location all that morning despite the loud purging process that a rocket goes through before a static fire test, blowing residual fuel from its lines. The engineers recovered the bird in a nearby field, and called an animal rehabilitation facility.

"These poor girls come out, and they're clearly feeling sorry for the owl, and they took him away," said Tom Mueller, who had watched as his engine roared to life. "He did not look good. I mean, the purges come on, that stand's louder than hell, but the owl was holding its ground, right? You'd think it would have flown out when the purges come on. No. But the engine lights, and he's like OK, I'm out of here."

They had come far, but SpaceX still had a ways to go before launching their rocket. Although the Falcon 1 on the Vandenberg test stand looked the part of a complete vehicle, in reality there was nothing in the second stage. It was an empty barrel. The second-stage engine, which would light and burn in the vacuum of space, was not ready. The rocket's avionics still needed work. And much more. But they had passed their first big test on the launchpad.

Chinnery and the rest of the SpaceXers were ecstatic, but also exhausted after months of working on unending technical issues. "We went out drinking that night, too, but it was a celebration instead of commiseration," she said. After the Friday night party, she went home and slept for the rest of the weekend.

While SpaceX's small team crashed into a euphoric slumber, senior Air Force officials realized they had a problem. Improbably, this brash-talking company had somehow built a rocket. They had test-fired it. And now, they were getting ready to launch it.

"Today we completed the largest milestone remaining before launch," Musk said after the successful test. "In a few months, we will receive Air Force clearance to fly."

Only, they wouldn't.

The Air Force and SpaceX had an uneasy relationship from the beginning. The military service has a rigid culture, strict hierarchy, and lots of requirements. SpaceX had a loose culture, almost no hierarchy, and viewed requirements mostly as a waste of time. SpaceX wanted to get things done, and the Air Force had people whose jobs it was to review every last environmental, safety, or technical detail before signing off.

For people like Koenigsmann, who had never really dealt with a big range like Vandenberg before, it was both amusing and frustrating. "The Air Force and us were such a mismatch in terms of how we talked, and expected things," he said. "They had some requirements that we were literally laughing so hard about. We would have to catch our breath. They probably laughed at us the same way."

But by early 2005, the company and the Air Force were not laughing anymore. Shotwell, who simultaneously was trying to sell the military

Falcon 1 launches and appease range safety officers, recounts one visit to Vandenberg that spring that left her uneasy. As she toured the site with the Air Force brass, Shotwell picked up a begrudging vibe from her escorts—all men. She felt it not so much from what they said, but rather the manner in which they spoke. "It was kind of like how I imagine a mafia meeting," she said. "They were acting like, 'You guys just can't do this.'"

Throughout 2004 and early 2005, the Air Force had nonetheless accommodated the rocket company. Chinnery, who understood the culture better than anyone at SpaceX, said she believes the Air Force leadership at the base simply didn't believe that SpaceX would succeed with its ambitious plans to build a rocket so quickly and launch it. They offered the minimum amount of support, and put their second- and third-string people on the job to look over paperwork and process approvals. But they were not obstructive.

As Chinnery put it, "There was not a whole lot of scrutiny early on. They just never really believed in it until, suddenly, the static fire happened, and they woke up."

The Air Force had good reason to doubt the promises of start-up rocket companies, as several had come through Vandenberg before. Their backers talked the same kind of talk that Musk did, about lowering the cost of access to space, offering a dedicated rocket for small satellites, and fundamentally changing the aerospace industry with newer technology and leaner operations. And inevitably, they fizzled out.

One of the most memorable had been the American Rocket Company, or Amroc, founded in 1985 by George Koopman. A colorful figure in Southern California, Koopman's interests spanned from Hollywood to space travel to the occult. He'd served as an intelligence analyst during the Vietnam War, made training films for the military, and coordinated stunts for Hollywood blockbusters. He was a friend to people such as Timothy Leary, who promoted the use of LSD for medical purposes, and the actor Dan Aykroyd. Through the latter connection, Koopman supervised stunts in the 1980 film *The Blues Brothers*, including obtaining permission from the Federal Aviation Administration to drop a Ford Pinto 1,500 feet from a helicopter into a small plaza in Chicago, surrounded by skyscrapers. He also claimed to have dated Carrie Fisher.

After founding Amroc in 1985, Koopman secured $20 million from investors and hired a team of engineers to develop an innovative hybrid rocket engine that uses both liquid-fuel propellants as well as solid, combustible materials. The engine's thrust of about seventy thousand pounds was on par with the engine that would power the Falcon 1 rocket. Koopman's speech at the International Space and Development Conference in 1989 sounds like Musk, a generation earlier.

"We want to target a reduction of 90 percent over the current launch costs," Koopman said. "We started a company to go into the business of transporting things to and from Earth orbit, a package delivery service. We wanted to be like Federal Express or UPS, and that is still precisely our objective." Like Musk, he wanted to make access to space routine, so that people could get on with doing business in space, and spreading the sphere of human activity far off of planet Earth.

James French, who had a distinguished two-decade career at NASA's Jet Propulsion Laboratory working on the Mariner, Viking, and Voyager missions, signed on as Amroc's chief engineer. He brought along a young Mike Griffin, the same engineer who would later become an early advisor to Musk, to work at Amroc. Griffin cashed in some of his savings and moved across the country for the job. For a time, the relationship was happy. But then they began to see Koopman more as a "wheeler-dealer" than a serious rocket scientist. Koopman had a good grasp of industry buzzwords, but his knowledge ran a mile wide and an inch deep.

Lacking government support and personal wealth, Koopman had to rely on funds from wealthy donors. In meetings with potential investors, Koopman would say his rocket was only six months from launching, and then put French on the spot to back his claims. The engineer tired quickly of this. "He'd make these outlandish claims, and he'd expect me to back him up," French said. "I could not, and we really got into some serious arguments about that." French and Griffin left after two years at Amroc.

Eventually, Koopman reached an agreement with the Air Force, and rebuilt an old launchpad at Vandenberg. Then, at the age of forty-four, Koopman died in a car accident in 1989. The company pressed on after the tragic loss of its charismatic founder, renaming the first test flight the "Koopman Express." In early October of that year, Amroc counted down to launch, but at liftoff the valve for a liquid-oxygen fuel line only partially opened, and with a limited flow of LOX, the vehicle did not have enough thrust to take off. The burning rocket toppled over onto the launchpad. Amroc suffered a similar fate, limping along for a few years before selling its intellectual property rights to SpaceDev, a subsidiary of Sierra Nevada Corporation.

When Elon Musk showed up at Vandenberg a decade later, some of the Air Force graybeards thought they knew pretty much what to expect from this private company. A lot of big talk about revolutionizing the space business. Fancy cars. And ultimately, a flameout.

The Air Force also had grand plans for Vandenberg in the early 2000s as it sought to rebuild the military's launch program. Decades earlier, under a deal brokered by the White House to help pay for NASA's space shuttle, President Jimmy Carter ordered the Air Force to fly all of its spy and communications satellites on the civilian space plane. The military had chafed when forced into a partnership with NASA for its launch needs, but dutifully began to phase out its old rockets.

On just its fourth launch, in June 1982, the space shuttle flew its first military payload into orbit. This forced marriage between civil and military space might have continued but for the space shuttle *Challenger* accident in 1986. This failure, beyond the human tragedy, allowed the military leadership to finally convince the White House that they needed their own, separate launch vehicles. Timely access to space, the generals argued, was impossible while NASA spent years investigating and addressing the shuttle's failure. And the military brass wanted modern rockets, not the converted ICBMs they'd flown on prior to the space shuttle era. The Reagan administration agreed, and the Air Force began to work with major defense contractors Lockheed Martin and Boeing to modernize their old Atlas and Delta rocket families.

By the turn of the century, the Air Force's long wait for modern, high-performing rockets neared its end. In 2003, Lockheed's sleek new Atlas V rocket entered the final phase of its development cycle, and needed a West Coast launchpad for polar missions. The Air Force assigned SLC-3 East to Lockheed for this purpose—the larger pad adjacent to where SpaceX would test its Falcon 1 rocket. Over the course of nearly two years, the Air Force invested more than $200 million to retrofit the existing mobile service and umbilical towers at the launch site, while also enlarging the flame trench. The Air Force had largely completed the makeover by the time SpaceX performed the Falcon 1 static fire test. And this was not the only valuable asset nearby. In the spring of 2005, just a couple of miles away at Vandenberg's SLC-4 site, a Titan IV rocket rolled out to the launchpad with a $1 billion spy satellite to be launched for the U.S. National Reconnaissance Office, or N.R.O.

As a result, when SpaceX met every requirement for permission to launch its Falcon 1 rocket, checking every box, the paperwork seemed to disappear into a black hole. The Air Force simply did not sign off on the final documents. For the Air Force, it came down to a simple calculation: let the new space company fly its unproven rocket, or protect its hugely valuable national security assets from debris or other hazards should the Falcon 1 launch go awry. The decision was an easy one for the generals.

The military officials would not permit a SpaceX launch until the Titan IV and its billion-dollar satellite for the N.R.O. lifted off. And they couldn't give a firm launch date for that mission.

The day after SpaceX's successful static fire test in late May, Musk and Buzza separately joined a teleconference with the Vandenberg base commander and the head of the N.R.O. The officials said SpaceX was welcome to launch its Falcon 1 rocket, but only after the N.R.O.'s expensive spy satellite was safely in orbit.

This placed SpaceX in a horrible position. While the Falcon 1 waited its turn, no one would compensate SpaceX for its expenses. The company got paid when it launched. By contrast, when the military awarded a national security launch contract to an Atlas or a Delta rocket, Lockheed and Boeing signed cost-plus agreements, where any delays were billed to the government, plus a fee.

"Technically, we weren't kicked out of Vandenberg," Musk said. "We were just put on ice. The Air Force never said no, but they never said yes. This went on for six months. The resources were draining out of the company. Effectively, it was just like being starved."

Almost from the beginning of its existence, SpaceX had pinned its hopes on this launch site, with easy access to polar orbits, located only 150 miles from its factory. In the haste to build ground systems, SpaceX had invested $7 million in launch facilities at Vandenberg. There would be no reimbursement. Musk had to eat the loss. He still had funds remaining from his initial investment, but with a payroll of more than one hundred employees SpaceX had maybe another year. And now he had to stand there and take it when the Air Force told him to wait, perhaps indefinitely, to fly from Vandenberg. The company born with DNA that impelled it to go fast, as fast as it could, had run into an immovable force.

It wasn't fair, but Musk had few options. The company faced a slow-moving bureaucracy. The Air Force had not said no. If they had, SpaceX could have fought the decision. But there was nothing to protest. A law-suit would not have brought an injunction against the military, and a favorable court ruling years later would have delivered a pointless, post-humous victory.

Knowing he could not wait, or sue, or protest, Musk took the only option left to him. After the conference call with the government officials, Musk phoned Buzza directly. We're going to Kwaj, he told Buzza. Tomor-row, he should begin packing.

Two years had passed since Chinnery's first visit to the Kwajalein Atoll in the middle of the Pacific Ocean, that surreal experience when Army officers had wined and dined her. Now the distant string of tiny islands five thousand miles away extended a critical lifeline to SpaceX. For the first half of 2005, Chinnery had lived mostly at Vandenberg Air Force Base. Now, she, Buzza, and a dozen other engineers and tech-nicians would spend almost the entire second half of the year in Kwaj, commuting daily by boat to Omelek Island. They had just worked them-selves to exhaustion building one launch site. Now they would have to turn around and build a second one.

Kwaj was far from home, but at least no Air Force officials waited there to shut SpaceX down. The Army wanted them. The range was theirs.

4

FLIGHT ONE

May 2005–June 2006

Surrounded by sparkling seas, Kwajalein is a tropical paradise. But it is an Army version of paradise. Instead of plush resorts, sweeping balcony vistas, and endless breakfast buffets, Kwaj offers two Army-run hotels with concrete walls, small windows, and a military-style cafeteria. The hotel named Macy's bears no resemblance to the department store, and the Kwaj Lodge is nothing special. Their drab rooms smell of mold, lack frills, and for entertainment offer an Army television set that picks up few familiar channels. Every piece of furniture has a U.S. government number stamped onto it.

"People either hated it, or loved it," Hans Koenigsmann explained. As for the laconic German engineer, well, he loved it. For although the Army had not dressed up paradise, it was no less a tropical island with abundant natural beauty. In his limited free time, Koenigsmann took advantage of the atoll's sublime diving opportunities.

One of his favorite haunts was an old German cruiser, the *Prinz Eugen*. Measuring seven hundred feet long, the World War II–era vessel rested upside down beneath 115 feet of water. The heavy cruiser had helped sink the Royal Navy's HMS *Hood* alongside the *Bismarck*, only to be surrendered to the Allies at the war's end. The U.S. Navy eventually dispatched the ship to Bikini Atoll, another part of the Marshall Islands, for atomic tests. It survived the Able and Baker tests in 1946, and was later towed into the Kwajalein lagoon and sunk. The ship's propeller and rudder protrude above the water line, but most of the ship lies below.

"You would go down there and hang out," Koenigsmann said. "We would swim down the keel, and then go under the ship, and up again. Because we worked during the day we ended up doing a lot of night diving. That's not for everyone."

The *Prinz Eugen* offered a reminder of Koenigsmann's roots in Germany, where the conflict's history and its consequences for the splintered nation remained vivid even decades after the war's end. After a comfortable childhood spent mostly in Frankfurt, Koenigsmann began to realize his talent for science and math. His favorite subject was physics, because it came easiest to him. He hoped to put those skills to use as a pilot. But his eyes weren't good enough. Aerospace, he thought, was the next best thing.

After a couple of years at a technical university in Berlin, Koenigsmann grew bored with airplanes. Then he discovered satellites and began working with other students to build small spacecraft that might one day go into orbit. In 1989, he moved to a scientific institution at the University of Bremen, in northern Germany. At this Center of Applied Space Technology and Microgravity, he led a team of five people that built a 140-pound satellite, named BremSat, to study tiny meteorites and dust particles near the Earth. As part of a NASA program to involve international partners in the space shuttle program, the agency selected BremSat to fly on a mission due to launch on *Discovery* in February 1994. During the year prior to launch, Koenigsmann made more than a dozen trips to the United States, visiting several NASA facilities, including Kennedy Space Center. He walked across the same platform the astronauts followed from the launch tower to the space shuttle. And, briefly, he met the mission's commander, four-time astronaut Charles Bolden.

After BremSat flew, Koenigsmann felt restless. He was nearing his thirtieth birthday, he had recently married, and his wife was pregnant. During his travels to America, he had liked what he had seen of the country. Together, the couple decided to make a leap. Koenigsmann moved his family to Los Angeles in 1996, taking a job at a small company named Microcosm. In California, he would help build lightweight rockets to make short hops. These "sounding" rockets lacked the punch to reach orbital velocity, but they could fly an arc that briefly rose above Earth's atmosphere and provide a few minutes of weightlessness to small payloads before falling downward. Koenigsmann didn't have much experience with rockets. None, actually. But he'd done a lot of hands-on work with controlling the flight of small satellites in space. Surely, he thought, it couldn't be that much different to do guidance, navigation, and control of a rocket.

"I thought it would be quite natural," he said. "It was part of the adventure. My wife really wanted to come over here. She just liked the idea of spending time in California." They would never leave. They got their adventure. And Hans Koenigsmann has since learned an awful lot about rockets.

Eventually, Koenigsmann's travels would take him further west still. Because he had accompanied Chinnery and Thompson on that first fact-finding trip to Kwaj, in 2003, he was already familiar with the terrain and the rigors of reaching the distant atoll. So when SpaceX activated its backup launch site in June 2005, he helped spearhead the westward charge.

They had so much to move. Unless an employee carried a key rocket part on the two-day, two-airplane journey from Los Angeles to Kwaj, most cargo had to be shipped by sea on a monthlong journey. At their headquarters in El Segundo, employees began to pack dozens of sea vans that would be stacked on large cargo ships for the voyage. Fresh from their experience building up Vandenberg, the SpaceX team knew they would need a lot of *everything* to assemble, test, and launch the Falcon 1 from Omelek Island. So they stuffed their tools and lifts and pipes and tubing and computers into forty-foot cargo containers and sent them to the Port of Los Angeles. Over the course of three months that summer, the company shipped about thirty tons across the Pacific, some by sea, and some by military air transport.

Koenigsmann and most of the others did not really mind the military-style accommodations because of a work schedule that called for early starts, long days, and full weeks. Before SpaceX acquired its own boat to go directly to Omelek, employees had to share the catamaran with the Meck Island passengers. This boat left the Kwaj dock before sunrise and would drop the SpaceX team off on Omelek about an hour later. They would work until late afternoon, when the catamaran would pick them up. In the evenings, back on Kwaj, there was plenty of planning to do for the next day: solving problems that had cropped up that day, conferencing with engineers back in California, plotting logistics, or working with Army officials to secure the necessary permissions to launch a rocket.

Some SpaceXers, like Chinnery, essentially moved to the atoll and spent almost the entire second half of 2005 on Kwajalein and Omelek. Others, like Koenigsmann, with his family back in Los Angeles, spent weeks on the island and then returned to the mainland. Over time, the back-and-forth proved grueling for the SpaceX workforce.

"I'd been to Hawaii zero times in my life," recalled Phil Kassouf, who was one of Koenigsmann's chief lieutenants in the avionics department during this period. "And in the span of six months, I'd been there so many times I never wanted to go again."

The company started with almost no infrastructure on Omelek Island, except for a small concrete bunker. The launch team had to pour a concrete launchpad and build a hangar to store the rocket. They bought massive, four-hundred-kilovolt-amp generators in California and shipped them to Kwaj for power. Brian Bjelde's first task was to make sure SpaceX could communicate with its rocket from Kwaj. Due to the explosive nature of rockets, and Omelek's small footprint, no one could remain on the island during a launch. So SpaceX set up its control center in an Army facility on Kwaj. From there, the range operators would need the capability to send a command to the Falcon 1 rocket if it went off course. Because Omelek lay over the horizon from Kwaj, Bjelde was not sure line-of-sight radio communications would work. On Omelek, he tried pinging Kwaj from the ground with a UHF antenna and other communications gear. The comms were too ratty. However, Bjelde then climbed into a JLG lift and rose to the point where he estimated the height of the antennas on a vertical Falcon 1 rocket would be. This time the signal went through just fine. A few dozen feet had made all the difference and saved the expense of building a station to relay signals from Omelek to Kwaj.

Bjelde and most of the other employees were also experiencing the tropical heat and humidity of life near the equator for the first time. During the first months there was no air-conditioning on Omelek, or any way to really cool off besides jumping into the lagoon. About the only place to catch a break on land was the concrete bunker, which was open at both ends. Inside, they could find some shade and moderate relief as the wind whistled through. Often, they would take their military-issue box lunches inside, eating their sandwiches, cookies, and bags of chips. They worked steadily, and by the fall of 2005 the pad infrastructure was mostly ready. It had taken only about four months to build a launch site in the middle of nowhere.

Chinnery attributes the speed to the learning experience of putting together the company's first launchpad, SLC-3 West, at Vandenberg, as well as more leniency from the Army officials at the remote location. This allowed the company to move at its natural pace: as fast as possible. "That's one thing SpaceX seemed to know how to do from the very beginning," she explained. "We just didn't waste any time dithering about stuff. If they knew they needed to ship stuff, they just shipped stuff."

When possible during the development of their Omelek launch site, SpaceX took the quick-and-dirty approach, sacrificing polish and sophistication for expediency. For example, the engineers decided they did not need a fancy "transporter" to move the rocket from the hangar, where it would be worked on and assembled, to the slab of concrete about 150 yards away where it would launch from. Instead, they devised a Bronze Age solution. The rocket would lay horizontal in a cradle, called the strong back, and this cradle had large metal casters as wheels that were designed to roll on a smooth surface. However, the terrain on Omelek consisted of packed coral, sand, and weeds. To cross it, the launch team would place large sheets of plywood on the ground, pushing the cradle five or six feet at a time, and then moving the plywood. Scrappy though it might have been, it got the job done. Once at the launchpad, the strong back would raise the Falcon 1 into a vertical position for launch.

Looking back on setting up the launchpad, Koenigsmann has a hard time fathoming how the company went so fast, especially because of the insane logistics of Kwaj and Omelek. "I know it's crazy, right? I don't know how we did that," he said.

The workload only increased as the first and second stages of the rocket arrived in September, introducing a whole new set of logistics nightmares. The launch team would be working on Omelek and a part would break, perhaps a pressurized dome, and they could go no further in the launch campaign without a new one. No one wanted to sit around for two weeks, far from home with nothing to do, so the engineers would fly back to Los Angeles while waiting for new components to get built, and delivered to Kwajalein.

Other times, a seemingly intractable issue would be resolved shortly after the launch team left the atoll. The SpaceXers carried Palm Treo smartphones at the time. On one occasion when they were flying home during the Flight One launch campaign, they had just landed in Honolulu when their phones started buzzing. The launch team were told they might have to turn around, but they were advised to continue home. Upon arriving at Los Angeles International Airport at 7:30 P.M., they learned they would in fact need to return to Kwaj, but the next outbound flight was not until early the following morning. After a few hours of sleep in their beds, at home, and a few words with family or friends, they were Kwaj-bound again.

That fall SpaceX moved a large trailer to Omelek, where engineers and technicians could sleep at night. This allowed for a small team to work on the rocket late into the evening, after the boats had left. These early accommodations were fairly crude on the remote island, so Chris Thompson had T-shirts designed for overnighters. At the time, the *Survivor* reality television show ranked among the most popular in America, so he used its logo as the basis for the design. But instead of "Outwit, Outplay, Outlast" for a motto like the TV show, the SpaceX T-shirts said, "Outsweat, Outdrink, Outlaunch." After surviving his or her first night, a SpaceX employee earned a T-shirt.

They had to work late because there was so much to do. As there were always problems to be dealt with, the engineers and technicians pushed themselves to exhaustion long after sunset. The propulsion team, led on Omelek by engineer Jeremy Hollman, had an especially difficult time. The Merlin engine's igniter system, which must reliably spark combustion between the rocket's LOX and kerosene fuel, provided an endless source of frustration.

Tempers escalated as the one to two dozen engineers and technicians on the Omelek team felt marginalized in the lead-up to Flight One. A few senior managers on Kwaj, often including Buzza, Thompson, and Koenigsmann, would consult with teams back at headquarters in California, holding teleconferences to troubleshoot issues. These instructions would then be called or emailed to Omelek in an attempt to assist their efforts. But at times this guidance felt heavy-handed to those working on the hardware.

This stress was compounded when the vice presidents on Kwaj began to complain about the lack of documentation as the first flight neared. Any work on the rocket needed to be carefully noted, the managers decreed. This irritated engineers on Omelek like Bulent Altan, who were doing their damnedest to get the Falcon 1 rocket ready to fly, while being pushed to go faster. They resented being asked to do things now that were not required before. One day close to the first launch, tensions boiled over. The vice presidents on Kwaj called to again complain about a lack of paperwork, forms, and tickets. "We got our asses chewed out, just this huge reprimand," Altan said. "We felt like slaves out on Omelek, with all the power stripped away from us."

And they were hungry. In the first year on Omelek, logistics were poor. Just as the team on the tiny island ran low on industrial supplies, sometimes they went without food, too. On the same day as the documentation blowup, the boat designated to deliver food, beer, and cigarettes did not show up.

"We had been going around the clock," Hollman explained. "We were sick of being told to do this, or do that. At some point everybody got fed up and decided that we needed to find a way to let them know that we were a part of this team as well."

And so they went on strike. Hollman put on his Telex headset and called over to the Falcon 1 launch director, Tim Buzza, back on Kwaj. The team on Omelek would work no longer without food and smokes, Hollman said. They had had enough.

Buzza recognized the gravity of the situation, and he hastily arranged for an Army helicopter to deliver a few trays of chicken wings and some cigarettes to the island that night. However, the pilot refused to put his helicopter down on Omelek, arguing that workers were in the process of erecting a tower on the launchpad that would carry fuel and power to the rocket during the countdown. It was unsafe. So Buzza improvised.

"I knew the helicopter pilot and promised him beers at the Snake Pit bar on Kwaj if he would do the supply drop," Buzza said. So instead of landing, they hovered above the island and dropped the food and cigarettes out the helicopter's side door.

Altan has another theory for why the helicopter pilot would not bring the chopper down. The launch tower was far from the helipad, he explained. But the bedraggled employees, a dozen of them or so, must have emerged from the dark—something like a scene out of *Lord of the Flies*. With tattered clothes, and white shirts covered in soot and kerosene grime, they swarmed the helipad when they saw the helicopter coming. "We were just wild animals on the island, waiting for food," he said. Ed Thomas, an ace technician who worked alongside Hollman, went immediately for the cigarettes. He stuffed two into his mouth, smoking them simultaneously.

With a little food and a little nicotine, the Omelek mutiny had been stilled.

Near the end of November 2005, the team felt ready. Working at a breathtaking pace, SpaceX had built two launchpads and a flight-ready rocket from scratch in three-and-a-half years. On Nov. 27, three days after Thanksgiving, SpaceX's pad crew arose hours before sunset to fuel the rocket with liquid oxygen and liquid kerosene. The Army had given the company a six-hour window, from midmorning to mid-afternoon, to complete the static fire test. Launch would follow a few days later. The countdown that morning was slowed by an unexpectedly cumbersome process of getting helium on board the rocket. (As propellant drains from fuel tanks during a launch, helium is used to push the remaining fuel and LOX into a rocket's engine.) To compound this issue, a problem occurred with one of the large LOX storage tanks on the island. One of its valves had been set to "vent" rather than close. SpaceX had to suspend the countdown and obtain permission from the Army for a handful of employees to boat over to the island and manually close the valve. Then they had to refuel the rocket, and ended up running out of time. During the countdown, the main engine's computer had also acted up. Musk told the team to work on both the fueling and computer issues, and try again in mid-December. They would have one more chance to fly in 2005.

SpaceX returned to make another attempt on December 20. Fueling operations were smoother, but this time the weather did not cooperate. Tropical winds blew across Kwajalein Atoll at speeds above 30 m.p.h., outside the safety levels set for a safe launch. Disappointed, the launch team began to offload the fuel to try another day. Then, disaster struck as they were emptying the kerosene fuel tank.

Thompson, who bore responsibility for the rocket's structure, watched the detanking process from the flight control room on Kwaj and saw something that did not quite seem right. "Wait," he said. "Is that a shadow?" Everyone looked up at the screen. The shadow continued to darken. Then, the tank separated from the rocket, and collapsed.

One of the pressurization valves had shorted out, and as propellant drained it rapidly created a vacuum inside the tank. The thin walls of the first stage began to buckle, collapsing inward. This threatened to destroy the entire rocket, engine and all. "It was just panic in the control room, as we were trying to basically stop the detank," Thompson said. With seconds to spare, they slowed the process of offloading the fuel, preventing an implosion that could have scattered pieces of the Falcon 1 around the launchpad.

Later that day, Thompson, Musk, Koenigsmann, and some of the others rode by boat out to Omelek to see the damage that had been done. The day's winds had kicked up waves across the atoll, and as the boat passed Bigej Island it struck a big roller, sending Thompson flying into the air. Musk recalls his chief of structures being flung several feet upward, coming down hard, and hitting the railing. By the time he got to Omelek, Thompson's knee had swollen to the size of a volleyball. The rocket was in bad shape, too. Its first-stage fuel tank had buckled completely, a total loss. As he was carried off the island back to Kwaj, Thompson knew the Falcon 1 would not be launching that year.

Despite these early setbacks, Koenigsmann had grown increasingly convinced SpaceX had taken the right approach to building a rocket. His experience at Microcosm in the 1990s provided a counter-example of how companies with less money and less urgency would fail. Microcosm's founder, James Wertz, did not have deep pockets, so he had sought to build the Scorpius family of rockets with dribs and drabs of funding from a succession of small government grants, beginning with a few million dollars from the Air Force Research Laboratory in 1993. The first step would be a small, suborbital booster, followed by a two-stage rocket capable of lifting a few tons to low-Earth orbit.

Microcosm's overall plans were not dissimilar to those of SpaceX—developing a simple, low-cost rocket that could be quickly rolled to the launchpad and flown into space. Twice, in 1999 and 2001, Koenigsmann helped Microcosm launch suborbital versions of its Scorpius rocket from White Sands Missile Range in New Mexico. The company characterized both tests as successful, but they really were anything but. Anne Chinnery, who worked for Microcosm at the time and helped facilitate range access at White Sands, said the 2001 launch of a larger Scorpius prototype lifted off but went off course fairly quickly. The problem was due, in part, to Koenigsmann's guidance system. He would take some important lessons from this failure when he left for SpaceX.

Koenigsmann learned the value of having enough money to do a proper job. As Wertz struggled to pay for booster development with small government contracts, the Scorpius program sputtered. Koenigsmann thought he might have found a solution for this problem after he first met Musk in early 2002, before the creation of SpaceX. An investment from the multimillionaire might invigorate Microcosm's rocket program. Eager to introduce Musk to Wertz, Koenigsmann arranged a meeting to discuss Scorpius and its funding needs. He hoped his boss and the investor would hit it off. But they did not.

"In my vision, Microcosm was limping along on their rocket program," Koenigsmann said. "And here's a guy who wants to build rockets. You put those two together, and there's a win-win for everyone, yeah? The problem is Jim had his own ideas on how to do this, and Elon had his own ideas, so basically Jim did not see that opportunity at all. I was livid."

A few weeks after the disastrous meeting, Musk called Koenigsmann. Would he consider coming and working for a new rocket company? He would. With his characteristic forward style, Musk arranged an interview *at* Koenigsmann's home in San Pedro, a community at the southern tip of Los Angeles that includes a chunk of the region's large port. At the time, Koenigsmann's parents were visiting from Germany, so he had to usher them out of the house and off to the movies.

The interview lasted for about two hours: one man born in South Africa, the other in Germany, sitting together in an American living room talking about space. "It's ingenious, actually," Koenigsmann said. "If you really want to find out about somebody, how they are, see them at their house. Check the kitchen and the bookshelf. I had a fair amount of technical books, and some classical stuff." The aerospace textbooks and Asimov novels would have impressed Musk, the Steinbecks not so much.

After six years in the United States, the Koenigsmann family had to decide whether their American adventure had run its course. If they were to stay—and mostly the family wanted to—Hans had to find a better job, with better pay and prospects. So it took about five milliseconds for Koenigsmann to say yes when Musk offered him a position. Musk had put $100 million of his own money into the project. There would be no limping along on skimpy government grants. The only contract point Koenigsmann sought to negotiate was an option for more vacation. He needed extra time to visit his family in Germany. Musk assented, knowing full well that Koenigsmann would be too busy to ever take that vacation. As indeed he was.

Early in 2006, after the painful implosion of the first-stage tank just before Christmas, SpaceX shipped a new stage to Omelek. Once it arrived, the launch team hustled to integrate this stage with the Merlin engine and prepare the rocket for a static fire test. The Army had given the company the first two weeks of February to complete this firing before it closed the Kwaj launch range for more than a month of maintenance. Any setbacks would cost SpaceX weeks of downtime.

During the first week of February, therefore, the launch team assembled the rocket and rolled it from the hangar on Omelek out to the launchpad. Once lifted into launch position, a dizzying number of lines had to be connected to both the first- and second-stage tanks. These lines supplied fuel, oxidizer, helium, and other gases and liquids to the booster. The rocket also required ample electricity to regulate tank pressures and open and close valves. All of this equipment had to be connected and disconnected every time the rocket was moved from a horizontal to a vertical position, and vice versa.

The cumbersome, nearly daylong task of connecting the second-stage fuel and electricity lines at the rocket's pointy end often fell to structures engineer Flo Li, who by then had earned the nickname "JLG Queen" for her skills on the lift, and Bulent Altan, one of Koenigsmann's other key deputies. Although the Falcon 1 was small by rocket standards, it still measured sixty-eight feet from the bottom of its engines to the top of its payload fairing, nearly the size of a six-story building. That was a long way to go in a small basket.

For Altan, who had struggled with a fear of heights for most of his life, it took courage to climb into the JLG basket beside Li. But those were his electrical cables running up a raceway along the spine of the rocket, and he bore responsibility for their wiring. And so, beneath the tropical sun, he would overcome his fears. "I would ride the JLG, white-knuckled, with Flo every time the rocket went up and down," he said.

Only a few days remained to complete the static fire test by the time Li and Altan had connected all the lines, in preparation for bringing the rocket to life. On February 6, the launch team attempted to power up the Falcon 1, always a tense moment because so much could go wrong. Due to the power needs on board the rocket, the avionics team had decided to crank up the voltage, pushing more electrical charge to the vehicle. They did so out of concern the rocket would suffer from a brownout if not enough current reached the vehicle along a lengthy cable. Despite this, when they tried to power it up, the rocket did not come alive that day. A power supply issue had cropped up with the second stage.

Li and Altan went back up into the JLG lift. They had a messy task to even reach the avionics bay inside the rocket's second stage. For the sake of simplicity, the door had no seal, but instead had been slathered with silicone to keep the interior watertight. Li and Altan ripped the sealant away, and then undid more than a dozen screws to get the door off. Almost immediately, the two engineers inhaled the ominous tang of burned electronics. They began to test various boxes that supplied power to components of the second stage, clearing them one by one. Finally, they came to one of the main power distribution boxes that Altan had built.

It had shorted out.

"That is when my stomach dropped," Altan said. "I knew it was my box and my design that had failed us, and I was going to cost us a month-and-a-half delay at least."

After checking schematics for the power box, Altan realized that the capacitors used in it were not rated to carry the increased voltage the launch team had pushed to power the rocket. The power box in the first stage also had the same capacitors, meaning it could also fail at any moment. It would be a relatively simple matter to replace capacitors for both of the power boxes, but rather inconveniently there were no electronics stores on Kwajalein. The capacitors needed for the job cost only about $5, but they had to be acquired from Digi-Key Electronics in Minnesota, six thousand miles away.

But these were desperate times, with just days left before the range closed. The launch team hastily devised a plan. An intern from Texas would fly on Musk's smaller Cessna Citation CJ2 from the SpaceX factory to Minnesota and pick up the capacitors. Meanwhile, as luck would have it, one of the thrice-weekly flights from Kwaj to Honolulu left later that day. If Altan scrambled, he could make that flight and arrive in Los Angeles on the afternoon of the following day, where he would meet the intern. Quickly, he and a technician removed the first- and second-stage power distribution boxes, and extricated the printed circuit boards inside. They were loaded into protective Pelican cases—Altan's only luggage. He raced in a boat back to Kwaj.

The flight into Honolulu landed at about 2 A.M. that night, with the next leg to Los Angeles scheduled for takeoff about five hours later. Such a layover seemed too short for a hotel, but the Honolulu airport closed its doors after the last flight of the night landed. Unable to sleep inside the terminal, a disheveled Altan found himself laying down on a concrete block just outside the entrance to the airport, trying to grab a few hours of sleep. The airport's closure did not stop the loop of recordings from playing above its entry doors, however. "I must have heard *Mahalo* over an announcement hundreds of times that night, and with all that combined with adrenaline, there was no sleeping," he said.

In Los Angeles, Altan's wife picked him up at the airport. She drove him straight to the SpaceX avionics building at 211 Nevada. The intern had already arrived with the new capacitors. It took less than an hour to change them out, and then an additional two hours to complete an "acceptance" test to determine that everything worked as intended. In the meantime, Altan returned home, changed into fresh clothes, and prepared for the second half of his trip. The return to Kwaj would be more comfortable, as Musk himself would be joining the team to supervise the static fire test. They all piled into Musk's larger jet, a Dassault Falcon 900, for the flight out to Kwaj. The Texas intern rode along as well, partly as a reward for his role in traveling to Minnesota, and in case another set of hands were needed.

Altan had hoped to sink into one of the private jet's spacious leather seats to catch up on missed sleep, but instead Musk peppered him with questions. What, exactly, had happened? How had broken electronics ended up on his rocket? Ever detail oriented, Musk wanted precise answers, and detailed plans for what to do when they arrived in Kwaj. Altan slept not a wink.

With its engines running, an Army helicopter awaited them at the Kwaj airfield. But first, as they always did, visitors to Kwaj had to fill out declaration forms to clear customs, and satisfy the Force Protection Bravo rules in effect at the Army base. This done, Altan and the intern hopped on the helicopter, Pelican cases in hand. They flew to Omelek and reversed course in installing and reconnecting the power distribution boxes inside the first and second stages. The vehicle powered up smoothly. Altan remembers remaining awake for a nearly sixty-hour stretch before he crashed on Omelek Island that night. The crazy capacitor odyssey had worked. The company would make its deadline, attempting a static fire test on February 13.

A less happy outcome awaited the intern, who still works in the industry and whose name I am withholding at his request. After learning that he would be joining the hurried flight to Kwaj, the intern inquired around at the factory about what the tropical experience would be like. The intern claims that his supervisor at SpaceX "recommended" he bring a small firearm, because there was a shooting range on the island. This explanation seems implausible, given that Kwaj was in fact a military base, and most everyone at SpaceX knew this. Regardless, the intern packed a handgun, and about one hundred rounds of ammunition. He dutifully declared this upon entry into Kwaj, but it had not been noticed by officials during the rushed customs procedures. The local police soon realized their error.

The Falcon 1 launch director, Tim Buzza, remembers standing in the SpaceX control room on Kwaj, monitoring activities on Omelek the day after the intern's arrival. "We get a knock on the door from the army police," Buzza said. "They asked if I knew the whereabouts of the intern."

He was dutifully ushered quickly back to Kwaj, where he and Buzza met with the island's provost marshal. While the intern was eventually allowed to return to Los Angeles, his career at SpaceX was over.

But that is not quite the end of the story. Buzza, Altan, and others remember well a "manifesto" of sorts that the intern sent to the entire company titled, "A farewell to my SpaceX family." In the email, he discussed his Southern roots, and sought to explain his actions and why he had felt compelled to bring a weapon to Kwaj. He also disclosed the name of his gun—Betsy. They would never hear from him again.

• • •

When the launch range reopened in March, SpaceX declared itself ready to fly the Falcon 1 rocket. The launch control team awoke early on Friday, March 24—if they rested at all in their rooms at the Kwaj Lodge, Macy's, and a handful of rented houses.

"I don't remember sleeping," said Kassouf, who by then stayed in a town house with Koenigsmann and Altan. "I was so excited, so nervous, so everything. It was like, oh my God, it's finally here. You have to keep in mind it was this marathon sprint. It always felt like we were a sprint away, but it just kept going."

Early that morning, they jumped on their bikes to ride toward the launch control center on the other side of the island. They pedaled into a headwind that invariably blew across the atoll in the morning, but they hardly felt it given the anticipation and adrenaline surging through their bodies.

Inside the spartan flight control room, Musk paced and fretted. He'd already flown out to the atoll on his private jet for a handful of static firings and launch attempts, and had grown eager to see the Falcon 1 fly. Later, he would temper public expectations before important launches, but he had not yet learned this lesson in 2006. Months before the Falcon 1 flight, Musk had told Jennifer Reingold, a reporter at *Fast Company*, that the Falcon 1 rocket had "well over" a 90 percent chance of success in its first launch.

As ever, Musk's mind bent toward the future. While Buzza and the launch conductor, Chris Thompson, worked through the countdown, Musk held a position at the back of the room on a platform. Throughout the countdown, Musk kept beckoning Thompson back to discuss materials for building the Falcon 5 rocket, his plan for a follow-up booster with five Merlin engines. Musk wanted more information on Thompson's plans to order a special aluminum alloy for the Falcon 5 fuel tanks. At about T—30 minutes, Musk walked to Thompson's console position and began a particularly heated conversation about why the materials had not yet been ordered.

"We were right smack in the middle of a count, and he just wanted to have this deep, aggressive conversation about materials," Thompson said. "I was absolutely dumbfounded that he was not even aware that we were trying to launch a rocket, and that I was the launch conductor, and responsible for basically calling out every single command that we're going to run. It just blew me away."

After Musk walked off, Buzza turned to Thompson and asked, "What in the hell is going on?"

In truth, this was just Musk being Musk, multitasking to the nth degree. Even in the middle of a critical countdown, he had the ability to simultaneously think about the company's needs six months or a year into the future. The last thing on Thompson's mind were shipping dates and aluminum costs. He had a rocket to launch. The company's *very first* rocket, in fact. Many of the things they were doing that day were new, and uncertain. But Musk's gaze looked far beyond the day's launch.

Despite the interruptions from the boss, the countdown proceeded more or less smoothly. And, to almost everyone's surprise, the clock hit T—0 without any kind of stop. The rocket's Merlin engine fired, and it started to rise. After more than a year of building up two launch sites, Chinnery watched from her vehicle control station. She could hardly believe her eyes as the rocket began to climb. "We finally counted down to zero," she said. "And when you've tried so many times, and it hasn't gone, you don't necessarily cheer right at zero. Because you're still half expecting the rocket to turn itself off and not go anywhere. So we waited a couple of seconds. And then you realize that it's actually flying away. It was incredibly euphoric."

Like almost everyone else in the control room, Musk's eyes were trained on a video feed of the launch. For five seconds, and then ten, the Falcon 1 rocket climbed above the sand, coral, and sea. Its flame burned brightly. It had actually launched. Nervous energy gave way to exhilaration.

Just as quickly, within seconds, everything started to go wrong.

Mueller was the first to notice a problem with the Merlin engine. "Oh shit!" he exclaimed.

Then everyone did. The engine itself seemed to be on fire.

"We noticed it on ascent," Musk said. "We were hoping that maybe if the rocket got far enough, then maybe the flame would extinguish, because it wouldn't have enough oxygen to burn."

The rocket did not get far enough for the atmosphere to thin. Half a minute after it took off for the first time, the Merlin engine flickered out. A few seconds later, the rocket itself stopped rising, succumbed to gravity, and fell back toward Omelek. Shocked launch controllers watched a feed from a camera at the launchpad that displayed burning chunks of the rocket falling into the ocean. An instant later, this camera feed blinked out. The Falcon 1's fire and fury had come to ground. For nearly four years, a small band of people had worked relentlessly to reach this moment. Within the span of one minute, it was over.

"It was amazing, and then it was horrifying," Chinnery said. "Something like that, it hits you in the gut."

The rocket blasted off at 10:30 A.M. local time in Kwajalein. After the failure, Musk and some of the other senior SpaceX leaders huddled with Army officials. Pretty quickly, they understood that a fuel leak near the top of the main engine had led to the fire. By then, it was shortly after noon, and it would take too long to assemble everyone, ride out in the big catamaran, and begin to collect the wreckage. At Kwajalein's latitude, the sun set before 7 P.M.

Musk and a few of his senior engineers decided to take a helicopter to the island that afternoon to survey the damage, before mounting a full recovery effort the next morning. As they flew over Omelek, they saw almost no visible debris. A parachute floated in the reef, but little else appeared disturbed. They began to put the pieces together of what must have happened. Saltwater spray coated most of the surfaces on the island. They saw bits of the rocket, but no large chunks. They realized the rocket must have crashed into the coral reef just to the east of the island. Later, another video of the launch confirmed this, showing the rocket exploding into shallow water, sending a torrent of ocean spraying across Omelek.

That evening, to raise the spirits of the launch team, Kimbal Musk decided to cook dinner. Elon's younger brother had been an early investor in SpaceX, and during the Flight One launch campaign provided public updates through postings on a blog titled Kwajalein Atoll and Rockets. After multiple stays at the Kwaj Lodge, Kimbal had tired of the lousy cafeteria food, so he ventured to the local grocery to peruse its limited options. As a chef, he made friends with a local military employee who had a small home with a backyard and an outdoor kitchen. After the Flight One failure, he cooked a bean-and-tomato stew, with some local meat, and a tomato-bread salad.

"It was large-format food, with no tables, so everyone just sat in a circle," he said. "It was a beautiful night, and a sad night."

In moments of high tension, Elon Musk often tries to break the stress with laughter. Musk has a rollicking wit. He will say something funny, realize it is funny, and iterate on the joke as a conversation proceeds. In this way, he brings those listening into the joke. As they sat outside, eating stew, Kimbal recalls his brother processing the day in his own way. Like other rocket enthusiasts, Musk experiences an extraordinary adrenaline rush during a launch. By the evening, he was crashing from that high, and beginning to reflect on what had happened, and what could be done to move forward. "He was clearly sad for what happened, but also making fun of the situation, because what else do you do?" Kimbal Musk said.

A pleasant surprise greeted the SpaceX team at the dock the next morning. More than one hundred people, mostly civilians working for the Army on the island, had gathered there. Kwajalein island's total population is only about one thousand people. These people weren't bound for Meck Island, but rather had come to show their support for the small rocket company. They wanted to jump-start the investigation by helping to pick up pieces of the Falcon 1.

Once the catamaran reached Omelek at low tide, the SpaceX team and civilians split up and combed the eight-acre island for pieces of the rocket. As part of the investigation, the military provided maps that searchers were supposed to mark where they collected pieces of the rocket. "It was kind of funny," Koenigsmann recalled. "We did this a little bit, but in the end it doesn't matter where this part falls. It came down because of something else that we knew about already. How it exploded, who cares?"

The rocket's small payload, FalconSAT-2, returned to Earth very nearly where it left, crashing intact through the roof of the machine shop, not far from its shipping container. Ironically, the forty-pound satellite waited half a decade to go into space, and at the end of its flight moved all of a few feet. Built by Air Force Academy students with just a $75,000 budget, the communications satellite had originally been scheduled to fly into space aboard NASA's space shuttle *Atlantis*. But NASA ended this program to orbit low-cost experiments after the fatal space shuttle *Columbia* disaster in February 2003.

The U.S. Defense Advanced Research Projects Agency, or D.A.R.P.A., stepped in shortly after the *Columbia* tragedy to offer another ride into space for the satellite. The defense agency had been looking for ways to support innovative launch concepts like the Falcon 1 rocket, so it bought the inaugural Falcon 1 launch. No one wanted to lose the satellite, but because it was not exactly an essential piece of national defense equipment, it provided a nice test payload.

"We were ecstatic," said Timothy Lawrence, then a lieutenant colonel overseeing the FalconSAT-2 project at the academy in Colorado Springs. "The satellite had hope. We were all on board from the beginning. We met SpaceX. It was a great relationship."

The launch agreement required a payload representative on Kwaj. But by the time SpaceX appeared set to launch the Falcon 1 for the first time in December 2005, all of Lawrence's students had gone home for the holidays. It fell to Lawrence, then, to make the trek to Kwaj. Due to flight delays, Lawrence was unable to make the last commercial flight from Honolulu to Kwaj, and he ended up on Musk's private jet. An Air Force lawyer said that would be fine, ethically, as long as the government reimbursed SpaceX with a fair commercial rate.

When Musk met Lawrence on his Falcon 900 jet before takeoff, he invited the Army officer to partake of scotch whiskey, Dom Pérignon champagne, and crab sandwiches. During the flight, Lawrence talked mostly to the pilots on board the aircraft, but he also watched the Musks carefully. While Kimbal played video games, his older brother spent much of the flight poring over books written about early rocket scientists and their efforts, such as the U.S. program under Wernher von Braun and the Soviet program under Sergei Korolev. Musk seemed intent to understand the mistakes they had made and learn from them. "I'm not surprised he has been successful," Lawrence said. "He was clearly dedicated."

Lawrence remembers watching the March 24 flight with senior Air Force leadership back in Colorado Springs, with live video provided by SpaceX. Mostly, he felt relief when the rocket launched. Although its fate was not a happy one, at least FalconSAT-2's saga finally reached a conclusion. The Smithsonian wanted to put FalconSAT-2 in the National Air and Space Museum, but Lawrence and other Air Force officers decided it might be best if the satellite served as a teaching tool for students. It can be seen today at the academy's museum.

. . .

On Kwaj, the Flight One salvage effort proceeded with only minimal coordination for a full day. Pieces were collected and brought to the integration hangar on the island. There, they were laid out from one end to the other, in the shape of the rocket. Pretty soon, a skeleton of the vehicle came into being on the floor. The day started out morbid, but as the hours passed, it became more of a fun scavenger hunt. One person might pop out of the water and say, "Hey, look, I found a turbopump!" After all, they realized, no one had been hurt. They would learn from this. And they'd get to orbit the next time.

Koenigsmann spent much of the day in the water. He found the parachute that had been stuffed into the upper end of the first stage of the rocket, about fifty feet of it. This was to be the company's first experiment in trying to recapture a rocket's first stage. The parachute, at least, had survived the impact and lolled in the water. He fought the waves, trying to haul in the whole parachute. But he could not pull it in. He grappled with the failure emotionally, too, taking it harder than most of his colleagues. Koenigsmann had led testing of the Falcon 1 flight computer and simulated every scenario he could think of. He'd game-planned all manner of possible failures, seeking to ensure that none of them actually happened to his rocket. And then it caught on fire and fell into the ocean.

"I was more of a failure-is-not-an-option kind of person, and this was supposed to work," he said. "So failure was definitely a hard lesson for me."

The loss of the inaugural Falcon 1 rocket dropped Koenigsmann into a depression. Later, his wife told him that he did not speak to her or any-one else for a month after returning to Los Angeles. He does not recall this, but said he must have just gone home in the evenings from work in a funk, lost in thought, trying to process everything. He was thinking about what had gone wrong, and what he could do to make sure it never happened again. She accepted that.

Musk seemed to recognize this emotional toll the failure might inflict on some of his engineers. Not long after the accident, he typed out an uplifting memo to the SpaceX team. He praised the performance of the rocket's main engine, its controlled flight, the avionics system, and more. After noting the preliminary diagnosis of the fuel leak, which began six seconds before liftoff, Musk wrote that the company would undertake a full analysis to determine exactly what had gone wrong. He hoped to try another launch within six months.

As part of his note, Musk also offered some comforting perspective. Other iconic rockets, he noted, had failed often during early test launches, including the venerable European Ariane fleet, the Russian Soyuz and Proton boosters, the American Pegasus, and even the early Atlas rockets.

"Having experienced first-hand how hard it is to reach orbit, I have a lot of respect for those that persevered to produce the vehicles that are mainstays of space launch today," he wrote. "SpaceX is in this for the long haul and, come hell or high water, we are going to make this work."

On their private flight back to California, along with several vice presidents and other company officials, the Musk brothers watched *Team America: World Police*, a 2004 movie that satirized America's role as global cops. Kimbal Musk said the movie's irreverence offered the perfect antidote to ease their tensions. "We watched it over and over again," he said. "There was no other movie that fit the mood."

Musk may have been able to laugh after the Falcon 1 failure, but he was not particularly amused. His rocket had blown up. How had it come to this, he wondered? And perhaps more important, who had screwed up?

5

SELLING ROCKETS

August 2002–August 2006

When he worked at Microcosm in the early 2000s, Koenigsmann made a number of good friends, perhaps none better than Gwynne Shotwell. Blonde and bold, Shotwell had plenty of brains, but none of the nerdiness or awkwardness that characterized some engineers. A former cheerleader in high school with a hearty laugh, she could talk to anyone. And often, she and Koenigsmann would go out to lunch.

So after the German engineer took a new job at SpaceX in May 2002, Shotwell celebrated by taking Koenigsmann to lunch at their favorite spot in El Segundo, a Belgian restaurant named Chef Hannes. Sometimes, to tease her friend, Shotwell called the eatery Chef Hans-y. After they ate, she dropped Koenigsmann off at 1310 East Grand a few blocks away. The large building was home to perhaps only half a dozen employees at the time. As they pulled up, Koenigsmann invited Shotwell inside to see his new digs.

"Just come in and meet Elon," he said.

The impromptu meeting might have lasted ten minutes, but during that time Shotwell came away impressed by Musk's knowledge of the aerospace business. He seemed no dabbler, flush with internet cash and bored after a big Silicon Valley score. Rather, he had diagnosed the industry's problems and identified a solution. Shotwell nodded along as Musk talked about his plans to bring down the cost of launch by building his own rocket engine and keeping development of other key components in-house. For Shotwell, who had worked for more than a decade in aerospace and knew well its lethargic pace, this made sense.

"He was compelling—*scary*, but compelling," Shotwell said. At some point during their brief discussion, she mentioned that the company should probably hire someone to sell the Falcon 1 rocket full-time. Jim Cantrell, who had not signed on as a full-time employee of SpaceX, worked sales as a consultant. At the end of the visit, Shotwell wished Koenigsmann well and left, hoping the new company would make it. Then she went back to her own busy life.

Later that afternoon Musk decided that he should, indeed, hire someone full-time. He created a vice president of sales position and encouraged Shotwell to apply. The prospect of a new job had not been on Shotwell's radar. After three years at Microcosm, using her mix of engineering and sales skills, she had grown the firm's space systems business by a factor of ten. She enjoyed her job. Moreover, by the summer of 2002, Shotwell felt like she needed some stability in her life. Unlike most of the recent college graduates Musk was hiring to work day and night, Shotwell had a lot to balance in her personal life. Almost forty years old, she was in the midst of a divorce, with two young children to care for and a new condo to renovate. It would be good for the aerospace industry to have someone like Musk come in and shake things up. But did she want to disrupt her life as well?

"It was a huge risk, and I almost decided not to go," she said. "I think I probably annoyed the hell out of Elon because it took me so long."

In the end, opportunity called, and she answered. Her final decision came down to a simple calculation: "Look, I'm in this business,"

Shotwell thought at the time. "And do I want this business to continue the way it is, or do I want it to go in the direction Elon wants to take it?" So she embraced both the challenge and the risk Musk offered her. After weeks of dithering about whether to stay or go, Shotwell finally called Musk while driving on a freeway through Los Angeles, toward Pasadena.

"Look, I've been a fucking idiot, and I'm going to take the job," she said.

Musk might not have realized it at the time, but he had just made arguably the company's most important hire.

Musk brought funding, engineering skill, leadership, and more to SpaceX. But to succeed in the global launch industry would require more than this. Aerospace companies in the United States, and institutional rocket businesses in Russia, Europe, and elsewhere, jealously guard their launch business. NASA, the U.S. Air Force, and other government agencies were generally comfortable with the existing state of things. And the large U.S. aerospace contractors had well-oiled congressional lobbies to ensure this order prevailed. To take all of this on, Musk needed a partner who possessed his brashness but also understood this political terrain and had the sophistication to navigate it. This was where Shotwell would come in.

She and Musk are both different and the same. He is blunt and, at times, awkward—she all smiles and smooth talk. But beneath their differing veneers they are sympatico, sharing the same fearless philosophy of charging forward headlong, seeking to mold the industry in their image.

Accepting Musk's job offer liberated Shotwell from the constraints of a more traditional aerospace company. During her first day at work, she set about formulating a strategy to sell the Falcon 1 rocket to both the U.S. government as well as small satellite customers. Seated in the cubicle farm at 1310 East Grand, Shotwell wrote a plan of action for sales. Musk took one look at it and told her that he did not care about plans. Just get on with the job.

"I was like, oh, OK, this is refreshing. I don't have to write up a damn plan," Shotwell recalled. Here was her first real taste of Musk's management style. Don't talk about doing things, just do things. She proceeded to generate a list of prior contacts in the industry, and people she thought might be interested in the small launch vehicle. Shotwell might not have had a rocket ready to launch, but she did have fortuitous timing. When Shotwell joined SpaceX in September 2002, the military had cause to be interested in what she was selling.

One year earlier, an aerospace engineer named Steven Walker was at his desk in the Pentagon when American Airlines Flight 77 crashed into the Department of Defense Headquarters. The 9/11 terrorist attacks left a strong impression on Walker, and the rest of the U.S. armed forces, as they scrambled to respond to a threat that originated in faraway Afghanistan. "One of the frustrations the defense establishment had is that, unless we had a base close to where the action was happening, it could take us a long time to intervene," Walker said. At the same time Shotwell joined SpaceX, Walker moved to the Defense Advanced Research Projects Agency to head a program created to address the military's need for rapid response.

Ironically, Walker's post-9/11 program would be named Falcon, for Force Application and Launch from CONtinental United States. (Walker was not aware of the Falcon 1 rocket when this name was chosen.) The Falcon program had two separate goals. The first involved development of a hypersonic weapon, and the second a low-cost launcher that could deliver at least one thousand pounds to orbit for $5 million per launch. In addition to giving the military a new capability, this would stimulate the stagnant U.S. aerospace industry. D.A.R.P.A. began to solicit bids from industry for the small rocket program in May 2003, and eventually received twenty-four responses. From these, Walker awarded nine grants worth about half a million dollars each for design studies. While some awards went to established companies such as Lockheed Martin, the majority were given to smaller firms like SpaceX. Ultimately, SpaceX and AirLaunch, which aimed to drop its rocket from a C-17 aircraft, emerged as finalists. AirLaunch never reached space.

D.A.R.P.A.'s support for SpaceX went beyond just this launch program. In 2005, the defense agency facilitated SpaceX's move to Kwajalein, working with Army officials there to ease access to the range, and helping to get the Falcon 1 rocket certified for launch. Walker also intervened with funding to move the Air Force Academy's FalconSAT-2 payload from the space shuttle program on to the Falcon 1 for its launch in 2006.

"In my mind it was less important that the satellite get into orbit, and more important that SpaceX and Elon get to a place where they could change the launch industry for the better," Walker said. "I was very impressed with their work ethic. There was some brashness there, of course. Some arrogance. But I think that comes with the territory. If you're trying to do something no other commercial company has ever done, you had better have some confidence."

For years after Sally Ride became the first woman to fly into space in 1983, she felt uncomfortable about serving as a role model for girls. Later in life, she explained in an interview how she came to peace with this position. "Young girls need to see role models in whatever careers they may choose, just so they can picture themselves doing those jobs someday," Ride said. "You can't be what you can't see."

Shotwell had a similar experience with engineering. In 1969, her father gathered five-year-old Gwynne and her siblings around the TV to watch the Apollo 11 landing on the Moon. She has fuzzy memories of the experience, recalling it as rather boring, and not as "good looking" as the childrens' shows she was familiar with. The rest of the Apollo program passed her by unheeded, never really sparking an interest in science. Growing up in Libertyville, a smallish town north of Chicago near the border with Wisconsin, Shotwell's life revolved around extra-curricular activities as well as work in the classroom. She captained the cheerleader squad, played varsity basketball, and enjoyed widespread popularity. But that would begin to change on a Saturday during her freshman or sophomore year. Some instinct prompted her mom to take Shotwell to a Society of Women Engineers event at the Illinois Institute of Technology. There, Shotwell soaked up career advice from a panel that included an electrical engineer, a chemical engineer, and a mechanical engineer.

"I loved the mechanical engineer," Shotwell said. "She was well spoken. She was incredibly poised. She had a beautiful suit, you've probably heard that, it's not a joke. I just thought she was great. Oh, and she was running her own business." The woman, in fact, owned a construction firm that focused on using green building materials, not exactly something in vogue during the late 1970s. "I fell in love with her, and I said I'll be her," Shotwell said. "And that's why I became an engineer."

As a senior in high school, Shotwell did not look far and wide for the best engineering schools. Of all the choices a straight-A student might have had, she applied only to nearby Northwestern University. She wanted a school that was strong in many nontechnical areas, not just engineering. When the prestigious Massachusetts Institute of Technology sent a letter encouraging her to apply, the name on the brochure turned her off. No way, she thought, would she attend a school named Massachusetts Institute of Technology. Shotwell had no desire to spend the next four years of her life considered a geek. "I wanted to make sure I was not a nerd," she said. "That mattered to me at the time. Now I celebrate my nerdiness, I celebrate my children's focus on engineering.

My husband's an engineer. My ex-husband is an engineer. His parents are both engineers. We revel in engineering now, but the world was a very different place."

College proved a difficult transition. Her freshman-year grades were marginal due to an active social life, and she struggled with engineering classes. A breakthrough came during a hard-core analysis class. Though she paid attention to the professor's lectures, the dense material seemed incomprehensible. But as Shotwell spent a weekend really trying to understand the fundamentals for the final exam, it suddenly began to make sense. When her professor handed back exams to the class, she had made the highest grade. It must have surprised the teacher, because when he returned Shotwell's test, he gave her a quizzical look. No doubt he wondered if she had somehow cheated her way to an A.

With her newfound confidence and improving grades, Shotwell began applying for a multitude of engineering jobs. On January 28, 1986, she had an interview with IBM. She had to walk through downtown Evanston to reach an on-campus interview, and paused to watch the space shuttle *Challenger* launching in a storefront television. With the first teacher to fly in space on board, Christa McAuliffe, the mission was big news across the country. As Shotwell looked on, with increasing horror, the vehicle broke apart seventy-three seconds into the flight while still in clear view of ground-based cameras. She soldiered on to the interview, not quite able to get past what she had just seen. "I was pretty shook up about it, actually," she said. "I didn't get an offer from IBM, so I must have really sucked eggs on the interview."

Her highest and best offer came from Chrysler, which hired a few dozen new graduates that year and paid them an annual salary of about $40,000, seeking to groom them for management. One week, Shotwell would find herself in a school for auto mechanics in downtown Detroit. "So me and the dudes were rebuilding engines, doing valve jobs, rebuilding transmissions," she said. "And I loved that." The next week she would work alongside company engineers designing new cars. Although she loved the garage work, the automotive engineering proved less than inspiring. A lot of the really difficult—and thus interesting—tasks were farmed out to contractors, often in foreign countries. So in 1988, after completing a graduate degree in applied mathematics, the Midwestern girl decided to move across the country for a career in a field that was still America-led: spaceflight. She took a job as a thermal analyst at the Aerospace Corporation in Los Angeles.

She got her first real taste of space in 1991, with the STS-39 space shuttle mission. The temperature of space changes rapidly when a spaceship goes from full sunlight into total darkness, such as when the shuttle would pass behind the Earth, opposite the sun. For this mission, the Department of Defense, NASA, and the international community flew several experiments on the shuttle, and when the vehicle opened its payload bay doors to space the "warm" payloads needed to stay warm, while the "cool" payloads remained cool. As a thermal analyst, Shotwell ran models of shuttle heating in real time as it orbited Earth, on supercomputers, and fed the data to Mission Control at the Johnson Space Center in Houston. This was fun, but after a while Shotwell realized that a company like the Aerospace Corporation, which did mostly analysis, might not be the best fit for her, either.

After a decade as an analyst, she joined Microcosm, focusing mostly on selling services to the government and space firms she had gotten to know at the Aerospace Corporation. During her three years at Microcosm, the company went from worrying about layoffs to expanding its staff. Yet this experience, too, still did not quite quench her thirst for making a difference. Deep inside, Shotwell knew she had more to offer the world. So the idea of selling Elon Musk's unproven rocket, and working for someone regarded as a demanding boss, did not faze her. "I knew the business by then," she said. "I would be selling to my old compatriots. Of course I could sell rockets. No question at all."

. . .

From the beginning, Shotwell understood the complex and evolving relationship between the Air Force, NASA, and private industry. Musk, however, was still learning about his new federal customers. In addition to selling rockets, part of Shotwell's job became managing relations between the company, her boss, and the U.S. government.

Shotwell and Musk spent a lot of time traveling together in those years. She and her ex-husband each kept their two children for a week at a time. On weeks with her kids, Shotwell would arrive early at SpaceX, and leave around 6 or 7 P.M., taking over for the children's nanny at home. During the other weeks, she could go crazy, working late into the night and traveling as much as needed. In 2003, she and Musk had traveled to Washington, D.C., for an introductory meeting with Peter Teets, then the director of the National Reconnaissance Office. Teets represented an important potential customer, as the agency designs, builds, and launches a multitude of spy satellites for the U.S. government. Teets was generally supportive of the start-up's intentions, but he had seen this kind of presentation before.

"I remember him putting his hand on Elon's back, almost hugging him, and saying, 'Son, this is much harder than you think it is. It's never going to work,'" Shotwell said. At this remark, Musk's back straightened, and he got this look in his eye that Shotwell could easily read. If Musk had harbored doubts about completing the Falcon 1 project before the meeting, Teets's paternalistic gesture had hardened his resolve. "You've just changed his mind," Shotwell thought about Teets, and the effect of his words on Musk. "He's going to make sure you regret the moment you said that."

While Shotwell met with potential customers, the rocket's design kept changing. During those first years engineers would plan, build, and test Falcon 1 hardware, and then change their design as parts failed.

The company's iterative design philosophy was new to many government officials working in aerospace, and accustomed to stable designs and slow-moving projects.

"Keeping up with that, and explaining to customers why we had a different approach to this or that, was a challenge," Shotwell said. Patiently, she would describe the company's method of making mistakes early in the design process, so it could weed those errors out of the final product. "These are government customers, so even though they wanted to move quickly, things changing as rapidly as they did still did not provide a lot of comfort. That was one of the hardest things I've had to work on for almost my entire career at SpaceX."

As Shotwell gained Musk's confidence, her role continued to expand. She managed customer interactions at first, but eventually added human resources, legal, and day-to-day operations of SpaceX to her portfolio. Her presence allowed Musk to focus where he could be most effective. On the days he works at SpaceX—which have varied widely over the years given his multitude of projects, but generally amount to half of a given week—Musk said he spends 80 to 90 percent of his time on engineering questions. This includes making design decisions, and optimizing the process by which SpaceX acquires parts from suppliers and builds its engines, rockets, and spacecraft. During meetings, Musk will make snap decisions. This is one of the keys that enables SpaceX to move so quickly.

"I make the spending decisions and the engineering decisions in one head," he said. "Normally those are at least two people. There's some engineering guy who's trying to convince a finance guy that this money should be spent. But the finance guy doesn't understand engineering, so he can't tell if this is a good way to spend money or not. Whereas I'm making the engineering decisions and spending decisions. So I know, already, that my brain trusts itself."

As SpaceX delved deeper into development of the Falcon 1 rocket in 2003, Musk came to believe customers like Teets needed to see real hardware to believe the company and its booster were legitimate. In the run-up to Thanksgiving that year, Musk pressed his new vice president of machining, Bob Reagan, to complete the company's first Falcon 1 rocket for public display. But the booster just was not ready, so Reagan's team had to mock one up. They worked eighteen-hour days to build the full-scale model, finishing on Thanksgiving Eve. The rocket itself was hollow, but from the outside the Merlin engine, and the first and second stages, looked genuine enough.

"To make a rocket look real is pretty hard to do," Reagan said. "We busted our asses. But in the end we made it look pretty bitchin'."

The day after Thanksgiving, the first Falcon 1 "rocket" rolled out of the SpaceX factory for a cross-country journey. Musk wanted to make a splash in the nation's capital, so his rocket was headed for Washington, D.C. It had almost arrived when, on the outskirts of the city, the tractor-trailer stopped as it crossed some railroad tracks. As it did this, the train signal's red lights and bells activated, and the crossing arm banged down on top of the Falcon 1. The damage was minimal, and the driver pulled up just before the train came through. When the truck reached the city, metropolitan police escorted it into the heart of downtown, where the sixty-eight-foot-long booster was parked flashily on Independence Avenue across the street from the National Air and Space Museum. If people needed to see a rocket, Musk was going to show them a rocket.

Shotwell admired Musk's talent for confronting his critics head-on, through showmanship like the Smithsonian event. Everyone thinks a private company like SpaceX building an orbital rocket is crazy, Musk told her at the time. They think it's never going to happen. "His attitude was, let's drive the damn thing there and show all of the detractors that it's here," she said.

The D.C. event took place on December 4, with temperatures just a degree or two above freezing during the evening hours. After Musk delivered a short speech standing in front of the rocket, he and other guests tucked into the museum for a reception of company employees, congressional staffers, and potential customers. Tim Mango, the lieutenant colonel helping SpaceX with its launch site in Kwaj, was among the invited guests. "That was probably the most interesting social event I'd ever been to," he said. "I was a paratrooper. A PBR guy. So to go to an event like that as a junior military officer, and being handed a glass of Napoleon brandy? That was something."

Musk's key lieutenants had come for the party. Mueller, Thompson, Koenigsmann, and Buzza dressed in tuxedos to escort their wives to the event and mingle among the crowd. As the festivities wound down, the vice presidents and their wives made plans to finish the night at a swanky late-night piano bar. But on their way out of the museum, Musk dragooned them into packing up the rocket. He decided it should be moved to a more remote spot for the night. So instead of singing along at a piano bar, the formally dressed engineers went back outside to the rocket, parked on the street. It was after midnight, with a light freezing rain falling, as they put a tarp over the Falcon 1 and prepared it for transport. Wet, cold, and tired, they returned to their wives after 1 A.M.

The Smithsonian stunt was the first forerunner to the "reveals" for which Musk, like Steve Jobs, would one day become known. In the case of the Falcon 1, Musk needed government customers to start placing orders for launches. This was how he believed SpaceX would one day become profitable. Beyond the small D.A.R.P.A. Falcon program, however, the government had not identified a small satellite launch need, nor had it issued contracts to build one. Rather, Musk anticipated such a need and self-funded development of a rocket to serve both commercial and government customers. He built the rocket on spec.

"When the government is hiring you to design, develop, build, and operate a thing, they're the customer," Shotwell said. "They're paying for it. They get to have their hands in the design. The decisions. They're covering the whole thing. But no one was paying us for design or development. They were paying us for flights." This offered an advantage in that SpaceX could build the rocket that Musk and his engineers wanted to—but it came with a big downside. Unless Shotwell sold a multitude of launch contracts, the company would die.

When he towed his Falcon 1 rocket into the nation's capital, Musk wanted to capture the attention of one government agency in particular. Following the space shuttle *Columbia* accident in early 2003, NASA had begun contemplating a postshuttle future. There were even whispers of using private spaceships to take astronauts into orbit, which could potentially save money and allow NASA to pivot toward deep space exploration. As the space agency began looking to the commercial space industry for help, it seemed to Musk that SpaceX had something to offer. But his hopes were misplaced. NASA had other plans, and this would lead to a pivotal clash between Musk and the nation's space agency. A clash that would ultimately save his rocket company.

Years before SpaceX came along, a U.S. company named Kistler Aerospace had also started working on a new rocket. Although both SpaceX and Kistler aspired to develop reusable launch systems, they could not have gone about it more differently. In 2003, as SpaceX feverishly built its first rocket, Kistler had already been at it for a decade without launching anything. The company had spent plenty of money, however. When Kistler filed for bankruptcy protection that year, it reported debts of about $600 million and assets of only around $6 million.

With his initial investment of $100 million, Musk knew he could not afford to build the Falcon 1 rocket if he relied on traditional suppliers for engines, software, and more. "Working with the giant aerospace contractors, geez, they don't even get out of bed for $10 million," he said. For this reason, SpaceX built as much of its Falcon 1 rocket in-house as possible, and sought out nontraditional aerospace suppliers. "Most of the aerospace people wouldn't even talk to us," he said. "Most of the time they didn't know who I was. And if they did know, I was some internet guy, so I was probably going to fail."

Musk taught his team to assess every part of the rocket with a discerning eye. Brian Bjelde remembers being constantly challenged. For a given task, a typical aerospace company would just use whatever part had always been used before. This saved engineers from the time-consuming, difficult work of qualifying a new part for spaceflight. The SpaceX attitude was different.

"True, a product may already exist," Bjelde said. "But is it optimized for your solution? Is it from a good supplier? And what about their tier two or tier three suppliers? And if you need more of them faster, will they meet your needs? If you want to change something, are they going to be willing to change it? And if you improve that product, will they then sell it to your competitors?"

Kistler, led by a cadre of former NASA engineers, approached its rocket design with a traditional mindset. The payload user's guide for the company's K-1 rocket, intended to launch from southern Australia and carry as much as four tons to low-Earth orbit, boasted about its blue-chip aerospace suppliers. "Each of Kistler Aerospace's contractors is a leader in its respective field of the aerospace industry and has significant experience in the construction of similar components," the guide stated. Among the contractors were Lockheed Martin (LOX tanks), Northrop Grumman (structures), Aerojet (engines), Draper (avionics), and so on. After seeking to integrate all these high-priced components, it is small wonder that Kistler found its financial situation dire by 2003.

But the company had a potential lifeline in NASA. Its longtime chief executive was a man named George Mueller, one of the heroes of NASA's Apollo program. He led NASA's human spaceflight program during the 1960s, and his management practices were widely credited for keeping the U.S. effort to land on the Moon on track. Later, he helped lay the groundwork for NASA's next major human spaceflight program, the space shuttle. After leaving NASA, Mueller became involved in several private ventures, and had led Kistler since 1995.

In February 2004, a year after the company's bankruptcy, NASA announced a $227 million contract with Kistler. The grant funded Kistler to finish building the K-1 rocket so the company could deliver supplies to the International Space Station. Some observers saw it as a gift for the Apollo legend, now eighty-five years old. The agency justified Kistler's award on the grounds that no other U.S. firm had a vehicle as near completion as the K-1 rocket. At the time, the company said it had completed 75 percent of the rocket's hardware, 85 percent of its design, and 100 percent of its flight software.

Musk seethed. He read the NASA news release and saw the very thing he detested. Here, he believed, NASA had used favoritism to award a contract, without competition, for something SpaceX might be able to do. And if Kistler launched cargo, it might one day be in line to launch humans into space as well. "Elon knew we were going to be building ships that were going to take people to the International Space Station," Shotwell said. "I don't know that I knew that at that period of time, but he knew that."

SpaceX, of course, had not launched in 2004, either. But it had started conducting regular engine tests. And it was about a year from a full static firing of its rocket. Musk wanted to protest NASA's award.

"I was told by many people that we should not protest," Musk said. "You've got a 90 percent chance that you're going to lose. You're going to make a potential customer angry. I'm like, it seems like 'right' is on our side here. It seems like this should go out for competition. And if we don't fight this then I think we're doomed, or our chances of success are dramatically lowered. NASA being one of the biggest customers of space launch would be cut off from us. I had to protest."

From the perspective of NASA and Kistler, the company had earned its cargo delivery award as part of a fair competition. A year earlier, Kistler won a separate NASA contract in a competition to demonstrate certain in-space capabilities. When the space agency began to evaluate potential providers for cargo supply to the space station in 2003, it determined that Kistler offered the best option. So NASA and Kistler renegotiated the terms of the earlier, competitively awarded contract.

"This was clearly creative on the part of NASA, but they had done their homework to evaluate other sources and Kistler was far and away a strong choice," said Rob Meyerson, who was then a program manager at Kistler, and would later serve as president of Blue Origin for fifteen years. "SpaceX was relatively unknown at this time."

Musk did not see it this way and would not be deterred. SpaceX protested. More than that, the company *won*. After NASA learned that the U.S. Government Accountability Office would rule in favor of SpaceX on the issue of fairness, NASA pulled the award to Kistler. The space agency realized it would need to open up a new competition for cargo delivery. This became the foundation for NASA's Commercial Orbital Transportation Services program, or COTS, that would emerge two years later and forever change SpaceX.

"I had nothing to do with the protest," Shotwell said. "It was all Elon. His vision carried the company."

The Kistler protest was just one of many battles that Musk and Shotwell would fight before government committees and in courtrooms. A year later, SpaceX scrapped with three of the titans of the U.S. launch industry. In one battle, SpaceX and Northrop Grumman traded lawsuits over Mueller and his rocket engine technology. And in another, more consequential battle, SpaceX sued Boeing and Lockheed Martin over plans to merge their launch businesses into a single rocket company, United Launch Alliance.

These two large government contractors had built many of the rockets launched from American soil since the dawn of the space age. Both undertook the development of modern rockets in the 1990s, with Boeing building the Delta IV line of rockets, and Lockheed its Atlas V booster. Accustomed as they were to sizable contracts from the U.S. Air Force and NASA for national security launches, neither company's new rockets could compete on price with rivals in Russia and Europe for commercial satellite launches. By 2005, the U.S. share of the global market for commercial launches, such as large satellites for television and other communications, fell to near zero. This left the two U.S. companies scrambling for Air Force contracts, alone, to make ends meet.

The competition between them for those contracts turned ugly, with allegations of theft. The U.S. Department of Justice investigated Boeing, asking how it acquired tens of thousands of pages of trade secrets belonging to Lockheed Martin. There were lawsuits. And then, concerned it might lose access to the Delta line of rockets, including the Delta IV Heavy, the U.S. Air Force stepped in. To end the legal hostilities the Department of Defense brokered a deal in which Lockheed and Boeing would merge their rocket building ventures into one company. Each parent retained a 50 percent stake in United Launch Alliance, which would be required to maintain both the Atlas and Delta fleets of vehicles ready for launch. For this, the company would receive a government payment worth about $1 billion a year, on top of individual launch contracts. The military got what it wanted—two independent means of reaching space. The big aerospace companies, Lockheed and Boeing, did, too. They now had a monopoly on national security launches for the next decade, along with guaranteed profits.

Everyone was happy—except for Elon Musk. He sued in U.S. District Court to stop the merger, arguing that SpaceX should be allowed to compete for these missions. While it was true that SpaceX had yet to launch a rocket, the company had big plans for a rocket larger than the Falcon 1. This was a violation of antitrust laws, he felt. This time, SpaceX lost.

"We weren't trying to break apart the old boys club; what we were trying to do was have fair access to competition," Shotwell said. "That's all it was about. Just be fair."

Almost from the beginning, then, SpaceX had to battle for its existence. One of the secrets of Musk and Shotwell's success is they did not kowtow to the existing order of large companies and government agencies. If they had to sue the government, they would. To fight back Musk would use everything at his disposal. Within its first three years, SpaceX had sued three of its biggest rivals in the launch industry, gone against the Air Force with the proposed United Launch Alliance merger, and protested a NASA contract. Elon Musk was not walking on eggshells on the way to orbit. He was breaking a lot of eggs.

This, of course, made life difficult for Shotwell, who would meet with competitors at space conferences and have to smooth ruffled relationships with government officials. Amid these controversies, about a year before the first Falcon 1 launch, Shotwell received an invitation to visit Lockheed's rocket production facility in Waterton Canyon, just south of Denver. The company had been building rockets there for half a century, since the early days of the Cold War. She does not recall why they invited her for a tour of the Atlas production factory, but Shotwell definitely felt that mafia vibe like what she had experienced at Vandenberg Air Force Base: *If you guys keep this up, we're going to crush you.*

What was most striking was the emptiness of the place. She visited in the middle of a weekday. It was not lunchtime. And in the vast factory no one was working on the rocket hardware. She said, "There were like three guys in this giant factory, with beautiful floors, all of this hardware, and no one's working."

Lockheed and other big launch companies had been reared on fat government contracts. For a given mission, they would receive its actual cost, plus a fee. The more something cost, generally, the larger fee the company received. The longer it took, the bigger the fee. One way

SpaceX differentiated itself, Shotwell said, is by pushing for fixed-price contracts, which incentivize a firm to get its work done. It also encourages the customer to keep the baseline request the same, and not make costly change orders to the design of a rocket or spacecraft.

SpaceX offered a consistent price for the Falcon 1 to both government and commercial customers—$6 million. It was remarkably low for the era, and it would be unprofitable until the company could launch dozens of Falcon 1 rockets a year. At the time, the only comparable U.S. rocket for the small satellite launch market was the Pegasus launch vehicle, built by a company called Orbital Sciences Corporation. Pegasus can lay claim to being the world's first privately developed orbital rocket, reaching space in 1990. Dropped from a converted airplane at forty thousand feet, the Pegasus rocket could loft a similar-sized payload as the Falcon 1 rocket. However, the rocket had a much simpler design, as it used solid rocket fuel (think fireworks) instead of liquids. It also used existing hardware, marrying state-of-the-art components into a new system.

This traditional approach, of course, led to higher costs. When Shotwell went to the market with the Falcon 1 in the early 2000s, a Pegasus launch cost between $26 and $28 million, so she could beat that handily—if her company could deliver. Because his price was so good, Musk wanted it front and center on the company's website.

This sort of transparency was pretty radical at the time. "It opened a curtain into a dark little corner," said Chad Anderson, who runs an investment group, Space Angels, that closely tracks public and private investment in spaceflight. "Before this there were a handful of companies serving the government and commercial launch needs, and it was more of a cartel situation."

SpaceX changed expectations with its low prices and transparency. Shotwell signed her first launch contract in 2003, to orbit a small experimental satellite named TacSat for the Department of Defense's Office of Force Transformation. This brought the company $3.5 million in revenue. The Malaysians paid about $6 million to launch the Razak-SAT satellite from Kwajalein. Including the payload for the first Falcon 1 flight, D.A.R.P.A. bought two missions. Along with some small grants, D.A.R.P.A ended up contributing about $16 million to SpaceX. These initial customers stuck with the company through its first launch failure.

"I think the early customers not only needed us to be successful, they wanted us to be successful, so they were going to hang in," Shotwell said. "Early customers don't hire maverick companies if they don't feel some kinship with the philosophy."

There's an old joke that goes: if you want to become a millionaire, start out as a billionaire, and found a rocket company. Musk was not a billionaire before he founded SpaceX, but four years into the venture it certainly had the look of a big money loser. Of the roughly $25 million in contracts SpaceX inked by the first flight of the Falcon 1, only part of that money was paid before a successful launch. This seemed like a poor return on a $100 million investment, and it's why NASA loomed as such an important potential customer.

In the spring of 2006, while Anne Chinnery, Tim Buzza, Hans Koenigsmann, and others toiled away on Kwaj to launch the first Falcon 1, Shotwell and her team in El Segundo had their eyes on a large financial prize. SpaceX wasn't running out of money quite yet, but the move to a new launch site and a growing payroll of 160 employees were taking their toll on the bottom line. Thanks to Musk's protest of the Kistler deal, however, NASA reopened the contracting process for using private rockets and spacecraft to deliver supplies to the International Space Station. This time, the agency received twenty-one bids for its COTS contracts, and in March 2006, NASA had whittled the applicants down to six finalists.

SpaceX stood among them. During negotiations with NASA that spring and summer, it became clear the space agency had two primary concerns: the quality of SpaceX's technology—the Falcon 1 rocket had just caught fire, after all—and whether Musk had the financial where-withal to complete the project. For SpaceX, meeting NASA's needs would involve not only scaling up the Falcon 1 rocket to a much larger launch vehicle, with nine Merlin engines, but also developing an orbital space-craft. No private company had ever done this before. The financial con-cerns were legitimate, too, as it seemed likely that Musk might need to bring in outside investors at the very least. Shotwell led a small team that responded to NASA's myriad questions.

As the summer of 2006 dragged on, SpaceX and the other finalists waited to find out who would receive the lucrative contracts to design their cargo delivery vehicles. The call from NASA finally came in August. Shotwell was upstairs, at the company headquarters in El Segundo, with Musk. After hanging up, they called an impromptu staff meeting out on the factory floor, where work briefly stopped on the next Falcon 1 rocket. Musk stood in the kitchen as the employees gathered around. His speech was short.

"Well," he said. "We fucking won."

SpaceX had won big in a couple of critical ways. First, there was the money. The contract value of $278 million would allow Musk to accel-erate his plans to build the big orbital rocket, and ensure the company's future while his team worked out its problems with the Falcon 1 vehi-cle. With the funds, SpaceX could also move into its larger, now iconic headquarters in Hawthorne. Perhaps most significant, with the contract award NASA had endorsed the company. "That was really important," Shotwell said. "We were a little company. We were jackasses at that time. We blew up a rocket in March of that year. From my perspective, NASA was acknowledging that even though we had a failure on Falcon 1, they felt like we had the right attitude."

In short, SpaceX had the right stuff. So did Gwynne Shotwell. Two decades after she had met that well-dressed, self-made mechanical engineer in Chicago, the woman who had convinced her that *she* could be an engineer, it was now Shotwell who wore the high heels and stylish clothes. And it was her time to show the good old boys how it was done.

6

FLIGHT TWO

March 2006–March 2007

Elon Musk recognized the extraordinary demands he placed on SpaceX's early hires. He therefore decided to reward employees who spent the majority of 2004 traveling to Texas for engine tests, and elsewhere. Anyone who spent two hundred days away from home in 2004 received an extra two weeks of time off in 2005, and an all-expenses-paid vacation wherever he or she desired to go.

"It was certainly a huge gesture," said Jeremy Hollman. "Only maybe ten or fifteen of us were eligible, and it was a testament to just how much we were all traveling and sacrificing." All Hollman needed to do was tell Mary Beth Brown where he wanted to go, and she took care of the rest.

Hollman planned to get married that year, so he chose to use the paid vacation for his honeymoon, a week in New Zealand and then a second week in Tahiti. Most employees who qualified took their paid holidays earlier in 2005. But Hollman waited until later in the year because he'd planned a fall wedding with his fiancée, Jenny. They picked a date in October when their families could all travel to her hometown of Mechanicville, a small town north of Albany, New York.

As work on the Falcon 1 progressed that fall, it seemed like the rocket might be ready to launch during Hollman's marriage ceremonies. Asked to delay his wedding, Hollman declined. Too many plans were laid, and he and Jenny had already sacrificed enough for SpaceX. They got married on October 8. Hollman had scheduled their honeymoon for later in the year, over Thanksgiving, thinking the first launch would surely occur by then. So when SpaceX finally made its first attempt to static-fire the Falcon 1 just after Thanksgiving, Hollman was absent.

"Jenny and I were in a bungalow in Bora Bora when SpaceX first tried to launch," he said. "We had really sparse internet and were only able to connect in the main lobby. I didn't find out until much later that it had scrubbed, and I didn't know why until I got back to L.A."

After returning from his honeymoon, Hollman immediately deployed on a multimonth stint to Kwaj for the Flight One campaign. Accordingly, he was on hand for the first launch in March 2006, as part of the fallback crew. The night before, he and a few other engineers and technicians, including Flo Li and Bulent Altan, had eaten a late dinner on Omelek, then stayed up under the stars, checking the launch equipment, and making jokes to ward off their nervousness. A few hours before sunrise they made the final preparations for launch, ensuring an adequate supply of LOX, checking the rocket one last time, and then retreating to Meck Island a few miles away before fueling operations began.

They watched from inside a concrete bunker, clustered together, keeping one eye on Hollman's laptop to watch data streaming back from the Falcon 1, and the other on Li's computer for video. After the Merlin engine failed, and the rocket tumbled toward the ocean, their excited shouts and chatter died. There was only one thing to do. After receiving clearance from the launch range, they scrambled outside, down to the dock, and caught a boat to Omelek. They rode mostly in silence, no one quite sure what to say. By the time they arrived to make a preliminary survey, the smoke had cleared. "I was in shock because, literally, we spent so much time putting that rocket together," Li said. "And then to see it that quickly break into pieces on the ground, it was shocking."

Not too long into their exploration of Omelek, the fallback crew heard the whirring of helicopter blades. It was Musk, Mueller, and several engineers from the launch control room to assess the situation, and begin collecting debris. In the aftermath of the failure, Li and the others appreciated Musk's encouraging words about finding the problem, fixing it, and moving on. "I always felt like he was super motivated to figure it out and go again," Li said.

Musk definitely wanted to go again. But he also wanted to reproach those who had let his company down. It is in his nature, after something goes wrong, to find whoever caused the problem and vent his frustration. In his mind, the Falcon 1 failure came down to sloppy work by propulsion employees who had been among the last to touch the rocket. He himself said so, publicly. Less than two weeks after the launch, speaking on April 5 at the prestigious National Space Symposium meeting, Musk said, "All current analysis shows that the nature of the problem was a pad processing error the day before the launch." The mistake had been made, he added, by "one of our most experienced technicians."

Within hours after the Falcon 1 launch, engineers traced the fault to a fuel leak. It appeared that a small B-nut had not been properly tightened on a kerosene fuel line. This simple nut provides a reliable seal at the connection of plumbing lines—if it is tightened to a specific torque. This particular B-nut had been removed several times during the run-up to launch to repeatedly access electrical components that were part of the relatively fragile avionics system near the engine. In fact, on the day before the March launch, Hollman and perhaps the company's best technician, Eddie Thomas, had removed and reattached this B-nut to access an igniter valve that needed rewiring. At six seconds before liftoff, when the fuel line opened, kerosene began leaking onto the engine at the location of the B-nut. When the engine ignited at T−3 seconds, the accumulating fuel lit into a visible fire. The rocket took off, but this fire forced the engine to shut down thirty-four seconds into the flight. Fire coming out of an engine's nozzle is good. Fire coming from the engine itself, however, is very, very bad.

After the Falcon 1 failure, Hollman was keen to recover as much rocket hardware as possible, study the data, and piece together precisely what had happened. As director of propulsion for the mission, he felt pride that the rocket had lifted off, and eagerness at the prospect of going all the way to orbit the next time out. Hollman passed a couple of days on Omelek helping to clean up, organize the debris, and batten down the site until the next launch attempt. This left little time for internet browsing until he settled into a commercial flight from Honolulu to Los Angeles. Tapping into the plane's rudimentary in-flight WiFi, Hollman remembers slowly loading news accounts of the failure. Eventually, he found some that suggested Musk blamed him and Thomas for failing to properly tighten the B-nut on the kerosene fuel line. It did not seem fair. There was no data to support the accusation. By the time his flight landed in Los Angeles, Hollman was pissed.

Hollman said he picked up his car and drove the two miles to SpaceX's factory in El Segundo. He parked out front, and said he marched into Musk's cubicle, ignoring protestations by Mary Beth Brown. Hollman did not like being singled out. But if he needed to, with his experience over the last four years at SpaceX, he could easily find a new job. But that might not be the case for Thomas. The technician had a daughter in middle school and a family to support. So Hollman raged at his boss for the public shaming and to protect the reputation of his valued technician.

After a few minutes, Hollman said, Gwynne Shotwell showed up and pulled him away from Musk's desk. Mueller arrived to meet with Hollman. For a while, they talked through the situation. Eventually, Hollman said he was going home for the weekend to cool off. He would come back on Monday. When he did return, Hollman told Mueller he had one condition for remaining at SpaceX: he would never have to speak with Musk again.

Musk does not remember events happening this way. "I have no recollection of Hollman storming into my cubicle area, nor would his opinion have held much sway with me," Musk said. "It would be accurate to say that his work, at times, did not meet the exacting standards needed for successful rocket flight." Musk also expressed goodwill toward Thomas, who remained with SpaceX for about a decade afterward. "I was at his retirement party and expressed my appreciation for his contributions in the strongest possible terms," Musk said.

Whatever exactly happened between Musk and Hollman, Mueller knew he did not want to lose his lieutenant, whom he had come to trust, and rely upon. Hollman had sweated in McGregor and Kwaj and been with the company almost since the beginning. Mueller made a stand. The Flight One mission manager, David Giger, said Mueller backed Hollman in the aftermath of the propulsion engineer's confrontation with Musk.

"He was basically like, if you don't want Jeremy at the company anymore, then you don't want me at the company anymore," Giger said. "I thought that was really awesome of Tom, and I think that's why Tom was able to build such an incredible team. He had their back."

In the end, Hollman and Thomas were not, in fact, to blame. The Flight One failure was caused by the tropical environment, not dereliction by Hollman and Thomas. The fuel inlet line in question was actually found among the debris on Omelek, with half of the cracked B-nut still attached by the safety wire that Hollman and Thomas had attached. When D.A.R.P.A. published its review of the Falcon 1 flight a few months later, the defense agency concluded, "The only plausible cause of the fire was the failure of an aluminum B-nut on the fuel pump inlet pressure transducer due to inter-granular corrosion cracking." The nut, which cost all of $5, had cracked due to corrosion from sea salt spray on Omelek the night before the first launch.

"That was just bad freaking luck," Mueller said. "The worst luck in the world."

Musk, Mueller, and Thompson had discussed the B-nut issue long before the first Falcon 1 flight. They had debated whether to go with aluminum or stainless steel. Most rockets Mueller knew of had used aluminum. Thompson, who had flown aboard helicopters in the Marines, said choppers sitting on U.S. aircraft carriers around the world exposed to sea spray used them. Ultimately, Musk greenlighted the use of aluminum B-nuts, because they weigh one-third as much as steel. With rockets, every ounce matters.

Prior to launch, SpaceX engineers had been concerned enough about the corrosive effects of sea salt spray to apply ACF-50, an anti-corrosion lubricant, onto parts of the rocket. But they hadn't been thorough enough in their application of the lubricant, nor had they accounted for the extent of harm caused by the corrosive environment. Even the vehicle hangar itself was not climate controlled. "Honestly, we did stupid things," Koenigsmann said. "We left the rocket outside for too long, and things like that. I don't think we appreciated, really, how harsh the environment was. We learned that lesson, too."

Perhaps most egregious, the launch team had left the rocket fully exposed on the launchpad for weeks at a time. On December 20, 2005, SpaceX had counted down to launch the Falcon 1 rocket, but ultimately had to abort the attempt. (This was when, afterward, the first-stage tank buckled during the process of off-loading the fuel.) At that point, only five days remained before Christmas, and the entire bedraggled launch team, who had sacrificed their personal lives for months, wanted to rush back to the mainland to salvage the remainder of the holiday season.

In their haste, they left the rocket outside instead of rolling it all the way back to the hangar. Made from fabric stretched over aluminum beams, the hangar was not the most sturdy structure, but it offered a modicum of protection. Taking the Falcon 1 rocket from a vertical to a horizontal position and returning it to the hangar would have meant that a good chunk of the SpaceX team would miss the last commercial flight home for the holidays. Buzza and the others figured that they would be able to return in early January, so the Falcon 1 rocket would remain exposed to the harsh environment for only two weeks at most. But because of delays in producing replacement hardware, they ended up not returning to the island and rolling the rocket back into its hangar until January 20.

The first-stage engine therefore spent an entire month outside in the merciless conditions. Trade winds blow across Omelek almost continually, carrying salt spray. The pad lies less than one hundred yards from the ocean, with big, breaking waves throwing salt into the air and coating the rocket.

"The corrosion environment there is insane," Musk said. "On Kwaj, it's like if you have a bicycle, there's always a prevailing salt spray. You have to keep your bicycle on the leeward side of the salt spray, or your bike is going to turn into a pile of aluminum oxide or iron oxide. The noobs on the island, they keep their bike on the windward side. Everyone who knows what's going on puts it on the leeward side, or their bike is going to be gone soon."

Corrosion or not, they would have lost the rocket anyway. When preparing the Falcon 1 for launch on Omelek, an engineer had opened a valve during the tanking process that allowed the second-stage LOX tank to vent more easily. This valve was never closed. Had the Falcon 1 rocket's first stage performed a nominal ascent, the second stage would not have remained sufficiently pressurized to boost FalconSAT-2 into orbit.

This all came out later in the data review. Musk asked his team why a computer had not checked for something like valve closures. The answer was simple: they did not have time to install sensors like this. SpaceX's first major failure, therefore, taught Musk there might be some limit to how fast a launch company should go. He still pushed, but he also gave his team room to work.

"As much as we thought we could launch quickly, we discovered that we had a number of shortcomings," Buzza said. "We just didn't want to fail on those for the next flight, because of how painful the first failure was. There was, from bottom to top at SpaceX, the idea that we needed to take some time and fix all these things."

Musk decided the second launch would carry a mass simulator, not a real satellite, so the company could focus on getting the Falcon 1 rocket right. And they would take the time needed, as nearly a full year passed between the first and second flights of the rocket. Gradually, throughout the rest of 2006, SpaceX began adopting more traditional aerospace practices. During a conventional rocket assembly process, someone meticulously records the serial number of every component or part added during the build. During construction of that first Falcon 1, no one had really kept those kinds of records. That would change for future Falcon 1 builds. Musk also got the sensors he wanted. The second Falcon 1 booster would carry all manner of devices to make certain that pressures, temperatures, and more were within acceptable limits. If someone left a valve open, the computer would flag it.

"In terms of our maturity and discipline, we were a completely different company coming out for Flight Two," Anne Chinnery said. "That failure helped us."

SpaceX also implemented design changes to improve the rocket, which they called Falcon 1.1. The company's engineers had gotten a lot smarter after spending a year building, testing, and flying the first rocket. For example, the avionics team learned better ways to route the miles of electronic wires throughout the rocket. From the outside the Falcon 1 appeared unchanged, but inside the guts were different.

More change came in the form of new leadership. Shortly before the Falcon 1 failure, Musk announced the hiring of the company's first president and chief operating officer, Jim Maser. The industry veteran had served as president of Sea Launch, a rocket company owned by four countries, since 2001. By the time Maser left, the company had launched nineteen Ukrainian rockets from a mobile sea platform.

Maser, forty-five, represented the classic, rite-of-passage hire for a start-up company—a few years after its founding a highly qualified, outside executive arrives to become the adult in the room. Such hires attempt to impose order upon chaos. "If your business grows, eventually you have to get out of the garage and become more professionally managed," Maser said. And so he tried to manage SpaceX more professionally. When he saw employees wearing flip-flops on the factory floor, Maser ended the practice. As a compromise for the unhappy workers, he consented to the continued wearing of shorts.

With two decades of Boeing aerospace heritage, Maser did indeed bring gravitas, and he helped institute some of the inventory control and quality inspection measures that spurred SpaceX's maturation between its first and second flights. But some employees, such as Koenigsmann, interpreted Maser's attitude as arrogance. To them, the new boss acted like he knew more about rockets than anyone at SpaceX. Koenigsmann had reasons for his frustrations. Maser was hard on the avionics chief, imposing rigorous qualification tests for the Falcon 1 computers that exceeded real-world conditions. As Maser started subjecting avionics components to more stringent tests, they began to break.

"I thought he should have known better, and I was very clear about that," Maser said of the German engineer. "And I think he disagreed with me."

Arrogant or not, Maser was certainly confident in his experience, and he sought to help SpaceX avoid mistakes previously learned by other rocket companies. Accustomed to working as a chief executive, Maser pushed for changes he thought necessary, and this led to clashes with the ultimate boss. After a few months at SpaceX, Maser told Musk the company should hire a couple of "systems" engineers to assess the rocket as a whole for potential risks. He also sought to implement more rigorous program scheduling. In 2006, Musk was already talking about Falcon 9 launch dates. Maser performed an independent assessment and found Musk's launch dates were far too optimistic. For Maser, these were things professionally managed companies should be doing. Musk viewed them as unnecessary impositions of bureaucracy.

"After running Sea Launch for so long, I was used to being an executive in charge," Maser said. "Ultimately, he wasn't willing to step back, and as I got more involved, we just started bumping heads. It became clear to me that Elon wasn't ready for me, and I was not the kind of person that would just follow orders."

Before the end of 2006, Maser left SpaceX to become president of engine maker Pratt & Whitney Rocketdyne. His tenure lasted nine months. It just had not been a good fit. Buzza said Maser brought a lot of strengths to SpaceX, but he was ultimately unwilling to be molded into Musk's image. "Chris Thompson also struggled some with that," Buzza said. "As far as working with Elon, I think Tom and Hans and I were able to walk that middle line." The middle line being that Musk listened to ideas. He encouraged debate. He empowered his senior employees funding and authority. But always, he had the final say.

Ahead of the Falcon 1 rocket's second launch, Musk called his vice presidents and senior engineers together to discuss their biggest concerns about the upcoming flight. Each of the company's primary divisions— structures, propulsion, and avionics—produced a list of its top ten risks for the mission. For example, a particular set of valves might have fared poorly in simulations, or a batch of components from a parts supplier might have failed qualification tests. The engineering team would discuss these problems and identify fixes.

Koenigsmann had a lot of problems to choose from. One issue he considered was a phenomenon called slosh in the second stage. When a rocket launches and burns propellant, its fuel tanks drain like water flushing in a toilet. As the tank empties, the remaining fuel can slosh around. If this effect becomes pronounced enough, it can cause the rocket to spin out of control. It is rather like running with a bowl of soup. If the rocket's movement couples with the slosh, the "soup" spills everywhere.

One means of controlling slosh involves inserting baffles along the edges of a fuel tank, which are basically metal plates that dampen the effect, and help orient the flow of fuel toward the upper-stage rocket engine. SpaceX did this for the first stage, but there is more of a mass penalty for doing so on the second stage, which rides all the way into orbit with the payload. So if they could be avoided, all the better. Because one cannot reliably test the performance of an upper stage in flight outside of an actual launch, Koenigsmann assigned Steve Davis to simulate fuel slosh with computer models.

Davis had been one of the company's early hires, joining in mid-2003, and by 2007 he managed Guidance, Navigation, and Control for the Falcon 1 rocket. He had three different models for slosh in the second stage, and ran simulation after simulation based on different assumptions. Most of the time, the second stage performed nominally. But in a tiny percentage, the rocket spun out of control. "It wasn't like just one number had to be off," he said. "A lot of things had to combine badly."

There were many risks with this second flight. When he prepared his PowerPoint presentation for Koenigsmann, Davis had about 15 risks for his systems. His top concern was actually a bending of the rocket in flight. Slosh ranked all the way down at number 11. "Slosh was a risk," Davis said. "But there were a lot of known risks heading into that flight."

When Musk came over to the avionics offices at 211 Nevada and met with the team to discuss their concerns before Flight Two, Koenigsmann presented his team's results. Ultimately, the team chose to accept most of the risks, including slosh. Addressing all the problems would involve months more of study, and potentially add significant weight to the rocket. And at SpaceX, the more direct solution was to simply fly the rocket, an acid test with more conclusive results than months of analysis, assumptions, and simulations.

They really could not afford to add more mass to the rocket, anyway. At the time, SpaceX was having difficulty building a rocket that could deliver its advertised one thousand pounds to low-Earth orbit. The Merlin engine's performance was not as high as the company anticipated, and it weighed more than they'd planned. Some parts of the rocket's mostly aluminum structure were also coming in heavier than predicted.

"As pretty much always happens, we were struggling with mass and performance," Musk said. "We were definitely in the red zone on payload, and every pound on the upper stage, we were losing pound for pound on the satellite."

When it came to adding slosh baffles, Musk also had concerns about the integrity of the second-stage structure. The baffles would have to be welded to the fuel tank's walls, which were made of an aluminum alloy. Because of mass concerns, these walls were already perilously thin for a stage holding fuels at high pressure. Welding slosh baffles would have introduced a weakness into the structure where they met the side of the tank. Such complexity, Musk thought, might well increase risk rather than reduce it inside the second stage.

"It wasn't even in the top ten risks," Musk said. "How bad could it be?"

As Musk and his leadership team debated the top risks for the upcoming flight, Flo Li and a few dozen other employees worked on the factory floor to complete the second Falcon 1 rocket. Li had been among the earliest hires at SpaceX, chasing her dream of going into space one day. On weekends, when she was younger, her parents would drive Li and her brother across the Chesapeake Bay Bridge from Delaware into Washington, D.C., to visit the National Air and Space Museum. One day, she found her destiny in the museum's IMAX theater, during a showing of the movie *Blue Planet*. Released in 1990, the film included arresting video of Earth from space, captured by astronauts on board the space shuttle.

"I was just like, 'Oh my God. That's what I want to do,'" Li recalled. She intended to go to space one day. And she would look down on Earth with her own eyes, marveling at the planet's blue beauty.

The childhood dream of becoming an astronaut stuck, and Li figured the straightest path to space lay through engineering. The University of Delaware didn't offer a degree in aerospace, so she studied mechanical engineering. She chose Stanford University for graduate school because of its reputation in aerospace engineering. And frankly, having spent the first two decades of her life in Delaware, Li also craved a bit of adventure. "I like to say Delaware made me a dreamer, because it was so boring," she said.

Li first heard about SpaceX in the spring of 2003, while preparing for a Ph.D. program she hoped would check a box on an astronaut application. She and some friends were drinking at a local dive bar, Antonio's Nut House, as was their custom on Thursday nights. The talk that night was of Musk, who had personally called one of her classmates to visit El Segundo for an interview. His vision of building a brand-new rocket stuck with Li, and during a job fair a month later, the SpaceX booth caught her eye. Li handed over a résumé, and soon she was invited to interview with Chris Thompson, and later Musk. SpaceX needed someone to help build the fuel tanks and exterior skin of the Falcon 1. Li had taken courses in structures but lacked hands-on experience. And she wasn't even sure she wanted the job. Stanford offered the prospect of a doctoral degree and a comfortable social life. SpaceX offered the certainty of very hard work.

She ended up taking the job, but immediately questioned her decision. While driving down Interstate 5 from the Bay Area to Los Angeles, her Volkswagen Beetle slowed to a crawl as the road descended into the San Fernando Valley. As she inched along, Li had plenty of time to think about what she had gotten into. Tears came as she thought of the friends she had left behind. She knew almost no one in Los Angeles.

But work soon swallowed Li's concerns. Who needed friends when there was no time for a social life? And anyway, she loved the people she worked, and soon bonded, with. She learned aerospace theory from Chris Thompson and other engineers, and then worked side by side with the technicians on the hardware when she joined the company in June 2003.

"I would leave work, and my head would just be spinning, just overloaded with information," she said. "I don't know when it happened, but after a few months I got into this groove where I just felt like this is all I want to focus on right now. It was like, this is my life. And I'm all in. And it felt really good because I just felt like I could put all of my attention and focus on what I had to do there."

Li steadily earned Thompson's confidence, becoming a key lieutenant in the structures department. Their overall mission was pretty simple. The main body of the vehicle had to survive the rigors of launch, endure high accelerations, and reliably contain volatile fuels under high pressure. And this structure had to be extremely light, or else the rocket would never go anywhere. Thompson referred to Li as "daughter number two," and delegated responsibility to her for the rocket's payload fairing. This is the cone at the top of the rocket that protects the payload during ascent into space. At a traditional aerospace company, a fairing would be purchased from a supplier. But Musk wanted SpaceX to design and build its own payload fairing.

As a starting point, she read a lot of NASA documents available on the internet. After choosing a design for the fairing, Li and a few other engineers built models to test their ideas. This hands-on engineering proved a far cry from the classroom environment, where she had learned about the theoretical properties of materials, such as strength and stiffness, and the mathematical basis for understanding the breaking point of a given structure. Now she was building things in the real world.

"In the early Falcon 1 days we did a little bit of everything," Li said. "I learned how to use a rivet gun, and how to weld things together. And then after we built something, we had to structurally qualify it. We had to convince ourselves, with testing, that this structure we checked on the computer, and that we built here, is not going to break when we launch into space."

A new first stage arrived on Omelek in November 2006, shifting the company's focus from the mainland United States back to Kwaj. At the beginning of the Flight Two campaign, Musk set a target of launching in January. Throughout December, the launch team checked out the first and second stages, and mated them together into a single rocket. However, as the launch team headed home for the Christmas holiday, with much work left to do, Buzza knew all hope of a January launch was slipping away.

Among the many challenges confronting the SpaceX engineers was getting the rocket to communicate with the ground support equipment. A couple of days after Christmas, Buzza and Koenigsmann were discussing an issue involving the Merlin engine's computer, which had intermittently dropped out during tests in December. They realized that to fix it, they needed to stop talking about it and work directly on the rocket. After convincing their wives it was the right thing to do, the vice presidents caught a flight to Kwaj. They would spend New Year's Eve and Day on Omelek, troubleshooting the issue and debugging software. For company on the lonely tropical island they had just themselves and the test site director, Sharon Hurst. The two men, both fathers and husbands, regretted sacrificing family time, but the work needed doing. They pushed through, drinking cheap beer and finding purpose amid the isolation.

"The remoteness of Omelek was, for me, almost inspiring," Koenigsmann said. "It was like another planet." It had been a difficult period. They had lost the first rocket. And here they were, once again far from their families. Yet it felt good to touch hardware again. For Buzza and Koenigsmann, 2007 dawned full of hope.

SpaceX missed the January launch date, but the engineers and technicians who returned to Omelek during the new year made steady progress. Before a launch attempt, the company would conduct two different major tests to ensure a rocket's readiness. The first test, called a "wet dress rehearsal," had nothing to do with clothing. Rather, the engineers would fuel the rocket, and then the countdown would proceed all the way down to the last sixty seconds or so. After the rocket passed this test, days or weeks later, the engineers would perform the prelaunch static fire test.

All of this testing required copious amounts of LOX. To supply liquid oxygen to the rocket, SpaceX ordered five-thousand-gallon containers shipped from the mainland. During the monthlong transit across mostly tropical waters, about one third of each super-chilled tank would boil off. They were ever running low on LOX. Amid a search for creative solutions, Buzza thought he had solved this problem by finding a machine advertised as capable of condensing air and separating liquid oxygen. Musk even signed off on the purchase order.

Phil Kassouf, the avionics engineer, remembers the machine arriving on Omelek in an unexpectedly huge package. Fully assembled, it measured about half the size of a standard intermodal shipping container. "It looked like it had come out of a mad scientist's laboratory, all spinning and whirring, with valves and gauges," he said. The engineers and technicians spent days greasing the machine, oiling it, lubing it, before finally turning it on. It belched. It made noise. And when the curious engineers returned after about forty-five minutes, Buzza exclaimed excitedly that the machine was working. It ended up producing about three hundred gallons a day, when not broken down, and supplemented the company's LOX supplies. But it was far from a silver bullet.

"That's the thing about Elon, he was willing to spend money to try things," Kassouf said. "And that's so different. Go to Boeing, and you spend money to try and figure out what your liabilities are going to be before you try anything. But Elon is like, sure, try it. If it doesn't work we can either sell it back, or it goes into our lessons-learned pile."

An unhappy fate awaited the LOX machine, which required a lot of power to operate. After a while, Buzza's team moved it back to Kwaj, where a Marshallese native named Jabwi was trained to operate it. He had to be inside when the machine whirred to life and churned out liquid oxygen. One night, Buzza was sitting on the porch at Macy's when the island's electricity blinked out. Sensing disaster, he immediately hopped onto his bike and rode over to the launch control center, where the machine had been deposited. The loss of power, he knew, would wreak havoc inside the complex machine, with sparks and fire. "It was a big electrical storm," Buzza said. "Jabwi thought it was the devil himself in there with him." The LOX machine ended up a charred ruin. Like the Army had done many times before, SpaceX dumped the machine into the lagoon. Today, it is an artificial reef.

During the campaign for Flight One, SpaceX labored for long, ago-nizing months in fits and starts to static-fire-test their rocket and launch it from Kwaj. By comparison, preparations for the second launch proceeded more linearly, over a matter of weeks, with new procedures and the benefit of hindsight. Things moved efficiently, that is, until the final forty-eight hours before liftoff.

SpaceX completed a successful static fire test of the Falcon 1 rocket's first stage on March 16, 2007. Just four days later, they were counting down toward launch. However, at T−60 seconds the flight computer automatically scrubbed the countdown after a pressure reading indicated a leaky fuel valve. It could be a real problem with the rocket, or spurious, like a bad sensor. There was no good way to know for sure without assessing the hardware. Buzza decided the rocket should be drained of fuel to address the problem.

Meanwhile, Musk watched the launch attempt from a Command Van back at the company's El Segundo headquarters. SpaceX had originally planned to use this converted tractor trailer as a mobile launch control center for missions from Vandenberg, but now it had been repurposed. With carpeted walls and TV monitors displaying live video from Omelek, Musk could track every step of the countdown. And now he was not happy.

Musk had waited very nearly a full year for a second attempt, and he wanted to launch. He pressed Buzza for why the rocket needed to be fully detanked to investigate the problem. Not safe, he was told. Could they just ignore the abort signal, reboot the rocket, and go for another try that day? Not if the sensor had identified a real issue that could threaten the mission's success.

Buzza was the launch director. It was his call. He ordered the rocket emptied of fuel. "Elon was super upset," Buzza said. "I suspect if he was in the control room on Kwaj he would have had his way, but having him five thousand miles away gave me a little wiggle room to take more time to figure out the issue."

After they detanked the rocket, the launch team dug in and discovered the issue was too complicated to be solved by a simple reset. Buzza's decision to scrub for the day was entirely justified. Over the following several hours, the engineers on Kwaj and Omelek solved the issues flagged by the flight computer. Buzza and his team left the control room in Kwaj at midnight to catch a few hours of sleep, feeling confident about the next day's launch attempt.

For the fallback crew on Omelek, a twenty-four-hour scrub meant another long night of readying the hardware for a second attempt. As night turned into day on March 21, Hollman, Li, and the others again worked on final preparations before fueling began. Unlike Flight One, however, Hollman and Li were not both watching the second mission from Meck. They would go their separate ways as employees evacuated the island. While Li took the boat with several others over to their bunker on Meck Island, Hollman rode in a helicopter back to Kwaj. For Flight Two, he would monitor the rocket's propulsion systems from SpaceX's flight control room.

The countdown proceeded smoothly the next morning, and this time the flight computer sailed right past T−60 seconds. The final seconds of the countdown crawled along agonizingly, so it seemed miraculous when the clock hit T−0. The engine ignited. There was smoke. There was fire. But then the rocket did not climb. During its very last systems check before liftoff, to ensure pressures in the combustion chamber were optimal, the flight computer found them too low. Another abort. A nightmare for the launch team.

"I knew, this time, I could not say no to Elon," Buzza said. "My brain was on fire. I needed to find another way out of this."

Buzza thought frantically. He conversed over his headset with Mueller and Musk, back in California, and Hollman, in the room. Together, they had spent years developing and testing the Merlin engine, and they knew it inside out. The sensor had read the chamber pressure at just 0.5 percent below the abort limit. This had been caused by slightly cooler than normal kerosene fuel, and this, they realized, was due to the previous day's failure. SpaceX stored kerosene fuel inside insulated tanks on Omelek at eighty degrees Fahrenheit. On the day before, when they had loaded kerosene from the tanks on the rocket, it had cooled down inside the first stage. Following the scrub, they offloaded the kerosene back into the insulated storage tanks. After less than twenty-four hours, however, sufficient time had not elapsed for the fuel to warm to expected levels. So when they reloaded the rocket the next day, the kerosene was abnormally cool.

Buzza and the others knew they had to slightly warm up the fuel in the rocket, which a sensor measured at sixty-four degrees. Buzza asked Hollman to calculate how much the fuel needed to warm to avoid triggering another automatic abort. The propulsion engineer came back with sixty-nine degrees Fahrenheit.

Five degrees warmer. That was all. Buzza determined that if they drained half the fuel on the rocket and retanked, the onboard fuel temperature should be seventy-two degrees. In the hour it would take to recycle the count and get back to T−0, this fuel could be expected to cool about three degrees. This would be cutting it close, but detanking any more fuel risked taking too much time, especially with storm clouds threatening on the horizon. Buzza shared his plan of action with Mueller and Musk back in El Segundo, and everyone quickly signed on.

At 1:10 P.M. on the island, the second Falcon 1 booster counted down for a third time. After the engine narrowly cleared the final chamber pressure test, the rocket blasted off. This time, the engine burned only where it was supposed to.

Inside the flight control room, Hollman did not pause to look up and watch the video screen showing the launch. Instead, his bleary eyes were glued to his monitor, which streamed data from the rocket's propulsion system. Arresting though images of a rising rocket were, the camera did not always tell the whole story. But data rarely lied. During Flight One, he had watched as the pressure inside the Merlin engine chamber fell to zero. Before most of the others, he knew the Falcon 1 rocket was coming back to Earth. But on Flight Two his data streams told a different story. Sensors in the thrust chamber reported good pressures. The rocket's temperatures were nominal. Fuel tank pressures were good. This time, the Falcon 1 rocket soared skyward. After a couple of minutes, the first stage fell away from the rocket's second stage, which continued climbing.

The Merlin engine had worked, by god. Hollman felt very, very good. Flight One had ended in bitterness, but Flight Two tasted much sweeter.

Miles away Li watched, too, with a growing sense of elation. Soon, the Falcon 1 crossed the threshold of space, soaring past sixty-two miles. Right on schedule, the payload fairing Li had helped to design, build, and test also fell away, chasing the first stage into the atmosphere. Staring at a video feed relayed back from the rocket, Li found herself watching a view of the Earth captured by something she had helped build. Suddenly she was a six-year-old girl again, sitting in a darkened theater. "Seeing the Earth from space linked back to my childhood memories, and it was so awesome," Li said.

It was awesome until it wasn't. A few minutes after the first stage separated, she noticed something was not quite right. The Falcon 1 rocket's second stage began spiraling off course. Slowly, it began to spin. As the spin accelerated to about sixty revolutions per minute, the Kestrel engine flamed out. The rocket had reached space, but it had fallen short of getting into a stable orbit around the Earth. The second stage began descending and splashed into the ocean a few hundred miles to the east of Kwaj, near a largely submerged spit of land called Kingman Reef. Later, when they realized this, they could all laugh a little bit—"Kingman" is *Koenigsmann* in English.

Despite this end result, with Flight Two SpaceX had come very close. Much closer than Flight One. Buzza said the control room on Kwaj felt pretty good about the launch. They had not achieved 100 percent mission success, but maybe 95 percent. They'd survived such a hard year since the first failure, and now they had gone through the entire first-stage burn, stage separation, second-stage ignition, and fairing separation. Less than half a decade after its founding, the company had climbed above Earth's atmosphere. SpaceX's next step was clear—orbit.

"We had always talked about needing three flights to get to orbit," Buzza said. "We actually kind of celebrated that night."

. . .

Musk, too, was feeling better about his rocket company. A few days after the flight, he said publicly that the mission represented a "large step forward" for SpaceX. This time, no one had to pick up pieces of the Falcon 1 rocket from the reef surrounding Omelek. Instead, Hollman, Li, and the other engineers cleaned up the launch site and took inventory so they would know what supplies were needed for their next attempt. After buttoning down Omelek, the team returned to California. They had a clear and attainable goal: reaching orbit on the next flight. They all felt confident about that.

Pretty quickly, it was clear that slosh had bitten them. As a handful of simulations had predicted, liquid oxygen in the upper-stage tank had begun sloshing around several minutes after launch, inducing a fatal oscillation. The problem they had known about, discussed in detail, and ultimately dismissed as the eleventh highest avionics risk took down their rocket.

"Now," Musk says, "I ask for the eleven top risks. Always go to eleven."

It's true. The company now does lists of the top eleven risks ahead of launches. That is one legacy of the second failure of the Falcon 1 rocket. Another is the never-ending tussle between a rocket's weight, the amount of payload it can loft to orbit, and the risk it will fail. And orbit is just the first, *easy* step. Going farther is more difficult. Musk wants to eventually go all the way to Mars. But for every pound of material a rocket can get to orbit, only a tiny fraction of that will reach the surface of Mars. Most of the rest of the mass is tied up in propellant, structures, and other hardware to push a spacecraft safely to Mars. So from the beginning, Musk understood he would continually have to make tough tradeoffs between mass, performance, cost, and risk.

During the early 1990s, in an effort to become more efficient and businesslike, NASA adopted a "Faster, Better, Cheaper" approach to space science missions. By the time SpaceX was founded, however, several high-profile NASA missions had employed this philosophy and failed. For any aerospace project, the joke became that you could never have all three, that a mission could never be faster, better, *and* cheaper. You had to pick two. But in pushing for high-performing, safe, and cheap rockets, Musk was not picking two. He was picking three. He wanted SpaceX to move fast, build better rockets, and sell them cheaply.

To build a better rocket, SpaceX had to limit its overall mass. So Musk fought to reduce weight. If he gave awards for rocket design, they would go to engineers who undesign things, those who remove mass. All too often, engineers want to add a part or a component just in case it might be needed during a contingency. Pretty soon, a rocket becomes fettered with widgets. Adding structure to a rocket induces a penalty, and that was why he decided against slosh baffles on Flight Two, accepting an ultimately fatal risk.

Musk also fought for efficiency. By mass, a rocket is about 85 to 90 percent propellant. So making an engine that requires even a tiny bit more propellant to produce a desired thrust can have huge mass consequences. Sometimes his team would seek to give up a bit of engine performance—known as specific impulse, or ISP—in their design. Just like a more fuel-efficient car goes farther on a tank of gasoline, a more efficient rocket engine produces more thrust. On suggestions like these, Musk pushed back *hard*.

"Inevitably with rockets, when you're trying to get into orbit, things look great in the beginning," he said. "There's lots of payload capacity. But then you lose a little performance here. Just give up a little bit of ISP there. Now you're a piece-of-shit rocket. You get chiseled away by 1 percent or 2 percent at a time. That's how it works."

Musk learned these lessons early on in the Falcon 1 flight program. His first rockets had nowhere near their advertised capacity of one thousand pounds or more to orbit. Several hundred pounds, more like. And if you're telling a customer you can get a half-ton payload into space, you either need to deliver or you lose the contract.

"That is why we fight for mass, and we fight for every fraction of a second of ISP," said Musk, who now wages this war with the Starship launch system, which he hopes will fulfill his goal of settling Mars. This ambitious spaceship sounds like science fiction. The monstrous first stage rocket has twenty-eight large Raptor engines. The second stage, Starship, may one day carry dozens of people to Mars and is designed to be reusable. Because of this it must set aside mass, already at a premium, for landing fuel.

"We fight for mass especially with a reusable upper stage, which nobody has ever succeeded in," he said. "Just FYI. It's not like other rocket scientists were huge idiots who wanted to throw their rockets away all the time. It's fucking hard to make something like this. One of the hardest engineering problems known to man is making a reusable orbital rocket. Nobody has succeeded. For a good reason. Our gravity is a bit heavy. On Mars this would be no problem. Moon, piece of cake. On Earth, fucking hard. Just barely possible. It's stupidly difficult to have a fully reusable orbital system. It would be one of the biggest breakthroughs in the history of humanity. That's why it's hard. Why does this hurt my brain? It's because of that. Really, we're just a bunch of monkeys. How did we even get this far? It beats me. We were swinging through the trees, eating bananas not long ago."

On Flight Two, Musk and SpaceX gambled by skipping slosh baffles, and it had cost them dearly in the end. But even monkeys can learn from their mistakes. For the remainder of the Falcon 1 rocket's flying days, SpaceX would install slosh baffles inside the second-stage fuel tank. Better to take a performance hit than not reach orbit at all.

7

TEXAS

January 2003–August 2008

Tom Mueller had watched Flight Two of the Falcon 1 rocket with white knuckles. Sitting beside Musk, the chief of propulsion felt his once-strong bond with the boss fraying. Although Musk publicly blamed Hollman and Thomas for the fuel leak that led to the first Falcon 1 failure, behind closed doors Mueller had not escaped Musk's ire. An uncomfortable rift had grown between the two men after they had worked so closely together in designing and building the Merlin rocket engine.

"My engine caught on fire, so I was in deep shit," Mueller said. "For the whole year between the first two launches, Elon and I were not good."

Following the first launch failure, Musk sought to buoy the spirits of his team by offering them a brief escape. At a cost of more than $100,000, he booked a private Zero-G flight to give employees a taste of spaceflight. Many had come to SpaceX with, if not the outright dream of becoming an astronaut, at least a hope that one day they might ride on top of one of the company's rockets. During the flight on board a 727 aircraft, about three dozen employees experienced several minutes of weightlessness as the plane flew parabolas, alternating between floating around the cabin at the top of the arc, and pulling nearly 2Gs at the bottom.

"It was like everybody who got a good grade got to fly on the Zero-G flight," Mueller said. "Well, I didn't get to fly in the Zero-G flight." Mueller, one of SpaceX's best engineers, had been left off the invite list.

Prior to Flight Two, Mueller and Musk had discussed the mission at length. Both agreed that once the second-stage engine lit, the rocket had it made. The Kestrel engine was simple and robust, and had never suffered any significant failures. When the Kestrel engine came alive on Flight Two, Mueller leaped out of his chair and cheered. Musk joined him, and they hugged and whooped together. In the span of a single moment, all was forgiven. Even the second stage's failure a few minutes later, as it spun out of control, could not entirely extinguish the mood. This one wasn't Mueller's fault. The rekindled trust was a good thing, as Mueller and Musk had a big task ahead of them. In California, Mueller and the propulsion team were designing a more advanced Merlin engine, which they would soon put on the test stand in Texas. This new engine would have huge implications for the company.

It would, in fact, very nearly break SpaceX.

A lot of rocket companies had been broken by then. A predecessor at Vandenberg Air Force Base seeking to build a low-cost rocket, Amroc, had gone bust in the 1990s. Some of SpaceX's earliest and most critical employees, including Koenigsmann, fled from Microcosm and its sputtering rocket program. In Mojave, where SpaceX performed its first gas generator tests in late 2002, a commercial space company named Rotary Rocket had run out of money only a year earlier. And the McGregor engine test site offered the most tangible mark of failure yet, with its immense tripod looming over the countryside—abandoned and silent as if it were the ruin of an ancient civilization.

The experience of Andy Beal and his Texas rocket company might have shaken the confidence of some entrepreneurs. Beal's launch venture certainly had not lacked for money. The Dallas banker regularly ranks among the world's two hundred richest people. He'd started Beal Aerospace with $200 million, twice the funding Musk put into SpaceX. And he had tried to do something similar—develop a big rocket that could serve commercial customers. He'd even had some technical success. By 2000, Beal had developed an incredibly large engine, the most powerful since the Saturn V rocket's main engine, and fired a prototype for twenty-one seconds. But Beal had encountered many of the same political and funding challenges that would bedevil SpaceX along its development curve. When Beal's company went under in 2000, he cited several reasons, including an inability to secure a launch site and NASA's favoritism toward traditional contractors.

"There will never be a private launch industry as long as NASA and the U.S. government choose and subsidize launch systems," Beal said in 2000, when he dissolved Beal Aerospace. "While Boeing and Lockheed are private entities, their launch systems and components are derivatives of various military initiatives." NASA, in other words, unfairly tilted the playing field against new launch companies.

In the wake of NASA's grant to Kistler in 2004, Musk would not have disagreed. But the founder of SpaceX was not content to issue angry statements or bow to the existing order. With a clear sense of what he thought was right and wrong, Musk took legal action when he believed NASA or the U.S. government paid unfair subsidies. So as he visited the McGregor site in November 2002, its history did not particularly bother him.

Once SpaceX leased the hundred-acre site, work proceeded quickly. Buzza, Allen, and a few employees set about pouring concrete, building a horizontal stand for initial Merlin engine tests, and restoring the blockhouse for workspace and test monitoring. The SpaceX team also began to understand the place, which in many ways remained wild, Texas ranch land.

Early on Musk brought his father, Errol, for a visit. The two have always had a complicated relationship, and Musk endured a difficult childhood. But he credits his father with teaching him the fundamentals of engineering. Musk did not realize it at the time, but as he built circuit boards and model airplanes as a kid, he was learning important lifelong lessons. "My dad is an extremely talented electrical and mechanical engineer," Musk said. "He tutored me, and I didn't even know it at the time." In 2003 the elder Musk lived in Los Angeles, and Elon thought he might be able to help with some construction work at McGregor.

As Allen took the two Musks around, they went into a building called the instrument bay, beneath what would become the Merlin engine test stand. Allen was tidying up the room as the others entered, but as he bent over to pick up one piece of paper, a diamondback rattlesnake hissed back. He returned the paper and calmly told the Musks to not approach the area. He walked out of the instrument bay, found a piece of steel, then came back inside and clubbed the rattlesnake. Errol Musk was evidently impressed. Allen heard him turn to Mueller and say, "You've hired that guy, right?"

There were other critters, too. In central Texas, black field crickets lay their eggs in the fall, and then these eggs hatch in the spring. About three months later, the crickets reach maturity, the adults get wings, and each one starts frantically looking for mates. As part of this messy process, thousands and thousands of crickets congregate into biblical swarms, which are particularly drawn to bright lights at night. They pile up like snowdrifts at doors and walls. According to Allen, the best way to kill them is not insecticide, but Dawn soap or liquid Tide.

"The soap suffocates them," Allen said. "It works better than any insecticide that we would ever use. But when they're dead, they stink like a dead horse."

They would pile up in great mounds. The engineers and technicians fought back with brooms and leaf blowers. But every year, the swarming crickets could rarely be kept at bay for long. At least they didn't bite. Black widow spiders are common in Central Texas, as well as the rattlesnakes.

None of this stopped the propulsion team from getting down to work. By March 2003, they had test-fired the engine's thrust chamber assembly for the first time, and drank their bottle of Rémy Martin cognac. Four days later, they prepared the thrust chamber assembly for a second, short test-firing. This night, the clouds were low, and the hour closer to midnight. The test was successful, and the team repaired to their apartments for the night.

When the white Hummer rolled up to the test site the following morning, the engineers had visitors. Two black Suburbans were parked at the front gate, waiting for the site operators to show up. Unbeknownst to the propulsion team, the horizontal test stand and business end of the Merlin engine pointed almost directly at President George W. Bush's ranch in nearby Crawford. Some very serious Secret Service agents wanted to know just what had happened the night before, when the test had shaken the ranch windows and awakened everyone. Bush was at Camp David that night preparing for the invasion of Iraq, but Secret Service agents remained resident at the ranch throughout his presidency. The agents asked a lot of pointed questions, and were not happy. Although SpaceX could not reorient the test stands, the company did gradually begin to get better about warning the surrounding community about future tests.

Mueller's team was able to move quickly into Merlin engine testing because of an early design decision. From the beginning, Musk wanted to build a reusable rocket. But the engine posed a problem. Rocket engines get really hot. Inside the Merlin combustion chamber, the flames from burning oxygen and kerosene can reach as high as 6,100 degrees Fahrenheit, and remain nearly as hot as the exhaust flows out of the chamber and through a nozzle. Located at the aft end of an engine, the nozzle shapes the flow of this superheated exhaust, allowing it to expand and accelerate. These temperatures are more than hot enough to melt aluminum, titanium, steel, or other metals commonly used in building engines.

One solution to avoid melting the engine is to cool its interior surface and nozzle. Much as coolant flows through an automobile engine to carry heat away, a rocket's "regenerative cooling" system routes room-temperature propellant through small channels in the engine walls to absorb the heat. This cooling system smartly uses the existing rocket fuel on board, but it adds complexity to the overall design of an engine. A simpler approach is to use an "ablative" material inside the thrust chamber and the engine's nozzle. As propellants combust, the ablative material chars, falling away in flakes, while protecting the chamber and nozzle beneath.

Prior to working at SpaceX, Mueller had considerably more experience with ablative designs. He worried that SpaceX would not be able to hire designers to plan an intricate cooling system for the Merlin engine chamber and nozzle. In some of their earliest discussions, Mueller convinced Musk that ablative engine chambers would get SpaceX to orbit faster. And the ablative design cost about half as much as a regeneratively cooled engine, he told Musk.

"He said the ablative nozzle would be a sure thing," Musk said. "Which it was not, by the way. That ablative nozzle caused us hell."

The ablative nozzle did, in fact, cause SpaceX all manner of hell in those early years at McGregor. Made of something akin to fiberglass, the ablative fabric is a resin mixed with silicon fibers. This "glass cloth" material is quite brittle to handle, and a small imperfection or slight crack introduced during its delicate curing process would lead to a much more substantial crack during testing. Desperate for chambers to test in late 2003, Mueller dispatched Hollman to the ablative chamber vendor, AAE Aerospace in Huntington Beach, to oversee their production. They could not keep up with demand as SpaceX began to test its engines for longer-duration firings.

The chambers cost about $30,000, and after taking delivery the propulsion team would perform a basic pressure test. With chamber after chamber, the ablative coating would bubble and then crack. Each failed chamber meant a delayed test in Texas, because if the propulsion team fired the Merlin engine for more than a few seconds, the ablative chamber would have to be replaced. The situation was dire. As Mueller put it, "The fate of SpaceX was kind of hanging on these chambers."

Then Musk had an idea. Perhaps, if they applied epoxy to the chambers, the sticky, glue-like material would seep into the cracks, and then cure, solving the problem. It was a Hail Mary. Mueller doubted the epoxy would stick to the ablative material, mixing about as well as oil and water. But sometimes Musk's crazy ideas worked, and he was the boss, after all. In late December, Musk had several failed chambers loaded onto his private jet and flown back to the SpaceX factory in El Segundo. When he met the team there, he was dressed for a Christmas party, wearing leather shoes, designer jeans, and a nice shirt. Late into the night, he and the propulsion team smeared epoxy onto the engine chamber. By the end of their efforts, Musk and the others were covered in the sticky stuff. He'd trashed his $2,000 pair of shoes and missed the party, but hardly seemed to notice.

This would be a sacrifice well worth making if Musk could save the Merlin propulsion program. And he believed he might have done just that right up until the moment the epoxy-coated engine chamber was subjected to a pressure test. It didn't take long, after pressures began to rise, for the epoxy to come undone. Soon, it flew off the interior walls of the chamber, revealing the cracks beneath. Musk had been wrong. But the filthy and exhausted engineers and technicians working with him all night did not begrudge Musk for keeping them at a task that proved fruitless. Rather, his willingness to jump into the fray, and get his hands dirty by their sides, won him admiration as a leader.

There would be no shortcut solution. Committed as they were to an ablative engine, the propulsion team continued tinkering with the design, and then testing potential solutions at McGregor. This was long, hot, dirty work. It would take months and a complete redesign of the compression wrap that supported the fragile ablative structure. This modified engine chamber and nozzle could withstand the brutal heat for a 160-second firing, but the changes made the engine chamber and nozzle thicker. It now weighed more, with lower performance, two things Musk despised.

The engine's combustion performance was measured by a variable called characteristic velocity, or C-star. After every engine test in Texas, Mueller or Hollman would have to call Musk and tell him the C-star value for that test. Higher values were better. With the ablative chamber, after months of toil they eventually reached very high values of C-star up to about 95, but they could only hold on to this high value for a few seconds before the engine's chamber blew. To reach orbit, the Merlin engine would have to burn for minutes. This meant the propulsion team had to detune the engine down to a lower C-star value, all the way back to an 87, where they had started nearly a year earlier. Every fraction lost meant that the rocket's payload capacity shrunk by that little bit more.

SpaceX installed a Panasonic video camera system to monitor engine tests in McGregor, which Musk frequently logged into from California. Sometimes, after a test, he would call first to get the test values. It would become a race for the Texas team to try and calculate the C-star value before Musk's call. Initially, the job of collecting data for this calculation fell to Hollman. After engine shutdown, he had to climb inside with a caliper to measure the diameter of the "throat" between the combustion chamber and nozzle. It was already hot in Texas, and to get faster measurements, Hollman would crawl into the engine before it had fully cooled. "This was by far the hottest and dirtiest part of the job," Hollman said.

After the data was collected and test results calculated, the call with Musk was often one that no one in Texas wanted to make or take. The company's chief engineer was growing increasingly disillusioned with the ablative engine design.

"You've got this real superheavy chamber with a crazy thick throat," Musk said. "So it sucks. It's heavy and your performance is weak. And then the irony is that it actually ended up being more expensive to make than a regen chamber, which is nuts. So now you've got a more expensive engine, which is single use, heavier, and more expensive. Ablative was for sure a huge mistake."

But at some point, they had to go fly. And once the propulsion team solved the ablative problem, they were stuck with it as the company sought the straightest path toward launch. The number one priority was to produce a flight-worthy engine—even if it initially sucked. In addition to the ablative issue, the propulsion team spent much of 2003 and 2004 fiddling with the engine's injector, determining precisely how much propellant to introduce into the combustion chamber at a given time, strengthening seals, and so forth. The problems never seemed to end. Initially, Buzza had thought that flying to and from Texas on Musk's jet was glamorous. But over time, the novelty wore off. It became grueling, especially for Mueller and Buzza, who had young children.

They lived dual lives. For ten days, they would work twelve- to fourteen-hour shifts in McGregor before flying back to California, where typically they would have Thursday through Sunday afternoon off. Then they would get back on Musk's jet for the return trip to Texas. For nearly two years, every other Sunday evening, Hollman would drive to Buzza's house in Seal Beach, picking him up on the way to the private airfield in Long Beach. Buzza's young daughters, Brandy and Abby, soon recognized the pattern. When one-year-old Abby saw Hollman coming up to the door, she would react by saying, "Jeremy bad." It made for painful goodbyes. "For years my younger daughter did not like Jeremy Hollman," Buzza said. "Because every time she saw Jeremy I disappeared for ten days."

He tried to make the best of a difficult situation. Often, before a trip to Texas, Buzza would buy two copies of a children's book, leaving one at home in California, and packing the other one. In Texas, Buzza would typically return late in the evening to his corporate apartment outside Waco. Fortunately, Texas was two hours ahead of California, so he could catch his daughters before they went to bed. He would call them, talk for a minute, and then ask his girls to find the book he had bought. Exhausted, sometimes he would wake up in the morning with the book on his face. The next night, one of his daughters would say, "Dad, you fell asleep again."

For all that, the work could be exhilarating; they were always chasing some new milestone. And after about two years, in January 2005, Mueller's team reached a major breakthrough with the first full Merlin test-firing. As the ablative material inside the engine chamber charred and flaked away, the engine kept burning. It ran until the propellant in its fuel tanks ran dry, shaking the bunker where Mueller and his propulsion team watched. SpaceX had just fired the Merlin rocket engine nonstop for 160 seconds, covering the full duration required for an orbital launch.

Still, they were not done. Although the Merlin engine had passed its most critical test, SpaceX still needed to push the Falcon 1 rocket's fuel tanks to ensure they could withstand the pressurization of launch and flight. This was especially true of the design chosen for the Falcon 1 interior. Its fuel tanks were stuck together like two beer cans end to end, with only a common "dome" between them. In most rockets, the fuel and oxidizer were separated into two completely segregated tanks. The Falcon 1 design saved mass but increased the risk, because only a single barrier separated the two propellants.

On the night of January 25, 2005, in Texas, Buzza greenlighted such a structural test. The engineers sought to determine whether the rocket's fuel tanks could withstand pressures greater than expected during a launch, hoping to get a sense of how far they could push the Falcon 1 without breaking it. They began to pressurize the tank, first to 100 percent of anticipated launch pressures and then, suddenly, at a pressurization of just 110 percent the rocket split in half. This was a catastrophe, as it meant they'd just blown up the first stage intended for Flight One.

Back in El Segundo, Musk watched the test unfold on the video link to Texas alongside Chris Thompson, his vice president of structures. They both were horrified. "The whole frickin' rocket just popped," Thompson said. "And the common dome is hanging out the side of the rocket, looking like a dangling radar dish. We're like, 'Holy shit, what just happened?'"

Musk and Thompson flew out to Texas that night to do a postmortem. The problem, they felt, was with the vehicle's welds. They were poorly done. The more they looked at the fuel tanks, the angrier Musk and Thompson got. A few years earlier, they had been impressed during their visit to Spincraft in Wisconsin, when Musk had burned his hands on the Holiday Inn Express toaster. But when Musk and Thompson flew to the company's headquarters in early 2005, they were no longer impressed. Walking into the Spincraft welding shop, Musk looked at the general manager, Dave Schmitz, and around the rest of the shop, Thompson remembers. Then Musk gave vent to his anger at the top of his lungs.

"You guys are fucking me and it doesn't fucking feel good," Musk bellowed. "And I don't like getting fucked."

The entire manufacturing facility ground to a stop. "You could hear a pin drop when he screamed that out. I mean, people stopped dead in their tracks, including all of us," Thompson said.

But it got the message across. By March of that year, SpaceX had a new first stage ready for testing at McGregor. Two months later, the Falcon 1 rocket aced its static fire test at Vandenberg. The rocket's first flight would follow in March 2006.

J ust weeks after the inaugural Falcon 1 flight, Tom Mueller had to make calls to prospective summer interns for the propulsion department. They needed additional hands at McGregor that summer. One internship seeker, Zachary Dunn, waited not so patiently in a Stanford University dormitory room. Musk's vision of low-cost access to space had entranced the onetime English major into moving across the country to chase his dream of working at SpaceX. In March, Dunn had watched the company's launch webcast from his dorm room, and it had fired him up. More than ever, he wanted to—had to—join the fight. Now he paced, waiting for a chance.

Dunn worried that maybe SpaceX didn't need someone like him anymore. SpaceX was no longer just a rebellious upstart with a few dozen employees and dreams of building a rocket. They'd built one, and they'd even launched it, however unsuccessfully. In Dunn's mind, he'd missed out on the most important, formative years of the company, and the best opportunity to make a meaningful contribution.

Finally, his telephone rang. Mueller. The propulsion chief began the call with a few technical questions about rocket engines. Nothing too difficult, Dunn remembers, just questions aimed at ensuring he knew the fundamentals of rocketry, such as understanding how a gas behaved in the intense environment of a rocket engine. And then Mueller asked Dunn if he would mind decamping to Texas for the summer. It was a long way away; the work would be hard and the weather shirt-soaking hot. Then, after only a few minutes, Mueller thanked Dunn for his time, and said he would let the student know at a later date if he had gotten the internship.

But Dunn was not ready to hang up like that. Perhaps Mueller intended to hire him, or maybe this was a polite brush-off. Either way, it was not in Dunn's nature to give up easily. He earnestly pressed his case, saying, "Mr. Mueller, this is my dream. This is exactly what I want to do with my life. If there's any question that you could ask me, anything I could do to demonstrate that I'm the right person for this, just let me know."

Having said his piece, Dunn paused and waited expectantly.

"OK," Mueller told Dunn. "You can come out to Texas this summer."

(SpaceX)

An overview of Omelek, the remote island on the Kwajalein Atoll in the Pacific that became SpaceX's launch site for the Falcon 1. Amidst a vast sea, Omelek is a tiny speck: it measures just eight acres, about the size of two New York City blocks. (Tim Buzza)

Chris Thompson, Hans Koenigsmann, and Anne Chinnery make their first visit to Omelek in 2003. The sun on Omelek could be relentless, penetrating T-shirts and all but the strongest sunscreens. (Hans Koenigsmann)

Omelek is only accessible by boat or helicopter, making the transport of both people and materials to the SpaceX launchpad a constant logistical challenge. Here, a Huey prepares to land. (Hans Koenigsmann)

Bigej Island, just south of Omelek, is where the SpaceX team would sometimes stop for a swim on their way home from work. (Hans Koenigsmann)

Elon Musk, founder and CEO of SpaceX, tours Cape Canaveral with President Barack Obama in 2010. (NASA/Bill Ingalls)

Gwynne Shotwell, right, was instrumental in making the fledgling company financially viable as vice president of sales. Here, she poses with NASA administrator Charlie Bolden, an important customer. (NASA/Jay Westcott)

Hans Koenigsmann, Musk's chief engineer of launch, on board a Huey helicopter two days before Flight Four. (Hans Koenigsmann)

Tom Mueller, a master of propulsion and architect of the Merlin and Kestrel engines, in his SpaceX cubicle. (Tom Mueller)

Tim Buzza, SpaceX's VP of launch and test. (Tim Buzza)

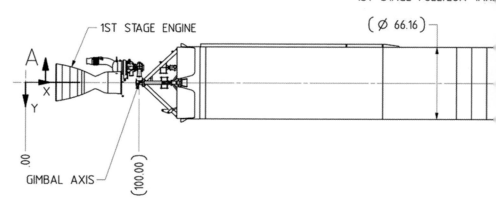

1ST STAGE ENGINE

1ST STAGE FUEL/LOX TANK

(Ø 66.16)

A

X

Y

.00

GIMBAL AXIS

(100.00)

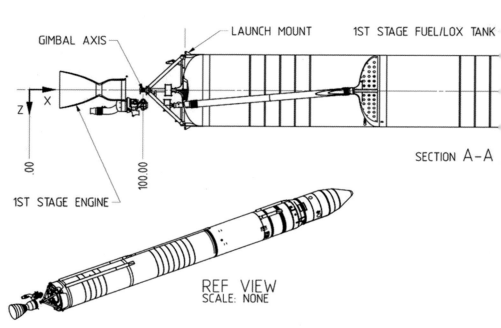

GIMBAL AXIS

LAUNCH MOUNT

1ST STAGE FUEL/LOX TANK

X

Z

.00

100.00

1ST STAGE ENGINE

SECTION A-A

REF VIEW
SCALE: NONE

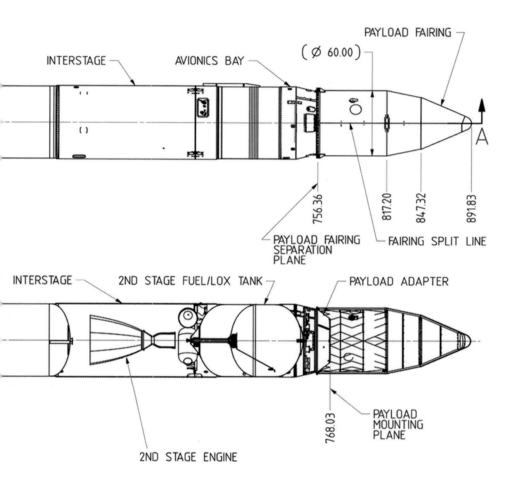

INTERSTAGE AVIONICS BAY PAYLOAD FAIRING
(⌀ 60.00)

756.36

PAYLOAD FAIRING
SEPARATION
PLANE

817.20 847.32 891.83

A

FAIRING SPLIT LINE

INTERSTAGE 2ND STAGE FUEL/LOX TANK PAYLOAD ADAPTER

2ND STAGE ENGINE

768.03

PAYLOAD
MOUNTING
PLANE

A Falcon 1 schematic from the rocket's payload user's guide.

Technician Ed Thomas with the second stage of the Falcon 1 rocket inside the hangar on Omelek. (Hans Koenigsmann)

A C-17 aircraft flies by Omelek Island during the Flight One campaign after delivering an emergency shipment of LOX. (Tim Buzza)

Elon Musk, center top, with the Flight One launch team and Air Force officials prior to launch. (Tim Buzza)

Collecting the wreckage after Flight One. (Hans Koenigsmann)

A solemn Elon Musk surveys debris collected after Flight One. (Hans Koenigsmann)

Kestrel engine with its expanded nozzle. (Hans Koenigsmann)

Zach Dunn poses with a Merlin 1C rocket engine on Omelek. Dunn, who started as an intern at SpaceX, would be instrumental in the development of Merlin. (Zach Dunn)

Tom Mueller, far right, celebrates the first Falcon 1 static fire at Vandenberg. To his left: Anne Chinnery, Dianne Molina, Bulent Altan, Phil Kassouf, and Jeremy Hollman. (Tom Mueller)

A Merlin rocket engine is test fired in McGregor, Texas. (Tim Buzza)

The launch of Falcon 1, Flight Two. (SpaceX)

Falcon 1, Flight Three. (SpaceX)

The hastily assembled payload for Flight Four, dubbed RatSat. (Hans Koenigsmann)

Launch team members pose with RatSat.
(Chris Thompson)

Short on time, SpaceX couldn't wait the month it would take to ship the Falcon 1 to Omelek by barge for Flight Four. Transporting a rocket by airplane is no easy task; here, the Falcon 1 first stage is wrapped for its flight on a C-17 aircraft. (Zach Dunn)

Brian Bjelde gives a thumbs-up next to the C-17 that would carry SpaceX's precious cargo for Flight Four. (Hans Koenigsmann)

Loading the first stage onto the C-17 for its journey across the Pacific. (Chris Thompson)

Crisis aboard the plane: Flo Li supports Zach Dunn as they work furiously after the Falcon 1 first stage implodes midflight. (Ron Gargiulo)

The moment of truth: Elon Musk and Tom Mueller watch Flight Four from the Command Van. (SpaceX)

It wasn't all hard work: Departing Omelek on the *Peregrine Falcon*, it's time to hang loose. (Hans Koenigsmann)

Zach Dunn with a shark he caught on Omelek. (Hans Koenigsmann)

Tina Hsu, left, and Flo Li hug a payload fairing. (Hans Koenigsmann)

Bulent Altan prepares his famed Turkish goulash. See page 267 for the recipe. (Bulent Altan)

Dunn had begun experimenting with rockets a decade earlier, in the rural, rolling hills of eastern Tennessee. He started with model rockets, but Homer Hickam's classic memoir *Rocket Boys* sparked a deeper interest in 1998. Dunn learned to grind the chemical potassium nitrate into a fine powder, then mix it with sugar. Once heated, the sugar would melt and coat the potassium nitrate, providing a simple fuel that could be packed into a pipe for launch. "I blew up a ton of rockets and hardware," Dunn recalled. "There were far more failures than successes." Eventually, however, some of his rockets rose about a mile into the sky.

On the eve of college, however, other influences pulled at Dunn. Interested in literature, he dropped a computer engineering major for English. Two years later he switched to geology, suddenly interested in volcanism. Perhaps that was *too* much science, however, as a year later Dunn had switched again, this time to mechanical engineering. Finally, the change stuck. By his senior year, Dunn was thinking a lot about spaceflight and the future of rocketry.

Like Musk had done four years earlier, Dunn searched for NASA's exploration plans on the space agency's website in 2005. By then, NASA had settled upon a plan to build Apollo-like rockets for an Apollo-like return to the Moon. The more Dunn read, the more he found himself rejecting NASA's plan, which seemed to rehash achievements of old. "I felt like the work that NASA was doing was not going to get us there," he said. Amid a phase of reading Ayn Rand novels, Dunn wondered whether, after years of public floundering, more nimble, focused private companies could do spaceflight better.

It was at this impressionable point that Dunn stumbled across SpaceX and its audacious founder for the first time. Old news accounts of Musk's winter visit to Washington, D.C., in late 2003—storming the very heart of government, with a shiny Falcon 1 rocket in tow—transfixed Dunn. This was the future. He knew what he must do. He would go and help Elon Musk build rockets to change the world.

Only by early 2005, SpaceX was nearly three years old. The company was already preparing for its first launch. Almost immediately, Dunn visited the SpaceX website to search for internships. He remembers the company advertising that it encouraged only the "absolute best of the best" to apply. He was just another mechanical engineering student. One of thousands across the country. What did they need him for?

The website informed interns that it sought go-getters who had excelled early and often. For example, the site suggested that successful applicants might have built a liquid-fueled rocket in high school. But Dunn's sugar-and-potassium nitrate contraptions had been relatively simple, solid-fueled rockets. "I didn't think I was ready," Dunn recalled.

In the end, Dunn applied for an internship at a different new space company that summer, Blue Origin, after graduating from Duke University. Although Bezos's space company was far more secretive than SpaceX, it shared the same basic philosophy—disrupting the aerospace industry by radically reducing the cost of getting people and stuff into orbit. Following a series of telephone interviews with the small company based near Seattle, Dunn received a call saying he had not made the cut, but that he stood as the "first alternate" for one of ten internship slots. If one of those students backed out, a recruiter from Blue Origin informed Dunn, the company would call.

"I told the recruiter on the phone, 'Just let me come out and work for you for a month,'" Dunn said. "You don't have to pay me anything, and if at the end of the month I don't work harder than every other intern, I'll just go home. It will be cool, and I will have appreciated the experience. But if I do come in and do an amazing job, and work harder than everyone else, then just pay me for the rest of the summer."

The recruiter rejected the suggestion. No one from the company ever called him again.

As he thought about his future career, Dunn did not know what step to take next. Perhaps, he figured, his best chance to work at SpaceX was to first earn a master's degree, and then a Ph.D. in rocket science, to really become a subject matter expert. He enrolled at Stanford University for this very purpose. Dunn made friends with a student group working on a hybrid rocket with elements of both solid and liquid propulsion. Dunn threw himself into this project, meeting an MBA candidate named Eric Romo who had worked at SpaceX from January 2003 through early 2004 on propulsion. In talking to Romo, Dunn realized that pursuing a doctorate in rocket science was the wrong approach. The company wanted doers, not scholars. He should get as much hands-on experience as possible while pursuing a master's degree, and do everything he could to earn an internship at SpaceX during the summer of 2006. As Dunn worked with Romo, the former SpaceX employee promised to put in a good word with Mueller, his former boss.

As soon as his spring classes ended in 2006, Dunn loaded up his Toyota Tacoma and drove for twenty-four hours straight, from California to Texas. In McGregor, he found a sleepy, one-intersection town typical of Texas, with lots of pickup trucks and a few fading storefronts. And a Dairy Queen, of course. Almost every small town in Texas has one of those. Upon reaching McGregor, Dunn passed through an industrial park before coming to a long driveway that led up to SpaceX's property, and toward the test stands and a blockhouse. Back in 2006, there was no guard house or gates to block one's entry. The only landmark visible from a distance was the tripod. Even so, Dunn approached the site in awe. "It felt like I had reached the big leagues," he said.

That summer, the company mood was upbeat. Musk hoped to launch the second Falcon 1 mission later that year, or early in 2007, and the propulsion team had lots of work to do. The second-stage engineers would soon bring a Kestrel engine to McGregor, and then the first stage was due to rotate through for extensive testing. It promised to be a busy summer with long, hot days and sleepless nights.

Dunn had about a week to acclimate himself to this new environment, and the test facilities, before a second-stage Kestrel engine arrived. In addition to the Texas wildlife, he and the others had to contend with summertime heat, which can become excruciating in Central Texas. The average high temperature in that part of the country reaches ninety-eight degrees in late July and early August. But during the summer of 2006, from July 12 through August 27, the mercury reached triple digits at McGregor on all but six days.

Dean Ono and other engineers from El Segundo who had helped build this Kestrel engine brought it for testing before the next Falcon flight. Once unboxed, the engineering teams worked around the clock on the hardware. "It was business time," Dunn recalled. "That was my first exposure to how intense the SpaceX work schedule could be."

By 2006, the propulsion team had grown. A typical day at McGregor would begin for some twenty engineers and technicians at around 8 A.M. and might last well past midnight. Once an engine arrived on-site, technicians performed some initial inspections in a small hangar to ensure nothing had gone awry during transport. Then they finalized the engine build with a few components, adding instrumentation so they could acquire rich channels of data during their tests. After this preparatory work, technicians would hoist the engine up onto a test stand. This raised platform held down the engine, provided conduits for rocket fuel to be loaded into the engine, carried away its exhaust, and provided myriad connections to receive data about the engine's performance.

Once on the stand, an engine first runs through a series of electrical checks, and then technicians ensure that critical valves that control the flow of rocket fuels open and close properly. There are more procedures, too, to make certain that none of the fuel lines or chambers in the engine leak, that nitrogen will purge all gases from the engine, and much more. All of this takes days.

The Merlin engine stands about ten feet tall, from the bottom of the bell to the top, where it interfaces with the rest of the rocket. Engineers and technicians worked in shorts, T-shirts, and tennis shoes, clambering over the test stand to work on the engine—straining to adjust this, or belly-crawling into a narrow space to fix that. All the while, classic rock blasted across the stands. The "Bear" radio station, KBRQ out of nearby Waco, transmitted one hundred thousand watts of Lynyrd Skynyrd, Aerosmith, the Rolling Stones, and more. For Dunn, despite the baking weather, these long days included sublime moments when he realized he was living the very dream he'd envisioned a year and a half earlier at Duke University. As he sweated, wrenched, and calculated, Dunn was in fact helping Musk build his rocket.

"Every day at about one P.M. the winds would blow up across the Texas test site," he said. "You'd be out there working on the test stand, and that felt fantastic. You had this rocket engine you'd be crawling around, rock 'n' roll music blasting, and starting to feel a breeze after a hot-ass morning. It was good living."

One of the trickiest parts of working with a rocket engine is lighting it safely. The liquid oxygen and kerosene that fuel the Falcon 1 rocket require an initial jolt of energy, or a spark, to begin the combustion process essential to produce large volumes of hot gas and propel the vehicle forward. This spark comes from an igniter. It sounds counterintuitive, but igniting a rocket engine consistently, and with precise timing, is enormously difficult. It had been the bane of the propulsion team's existence in El Segundo, Kwaj, and McGregor for years.

To light its Falcon rockets, SpaceX first used hydrogen igniters, and later switched to a volatile chemical mixture known as TEA-TEB to ignite the Merlin engine. This is a combination of triethylaluminum (TEA) and triethylborane (TEB), which are essentially two different metal elements each linked to three hydrocarbon atoms. These molecules are held together by rather tenuous bonds that will break easily. In fact, when TEA-TEB comes into contact with oxygen, it spontaneously combusts, producing a green flame. So to start a rocket engine, oxygen is pumped into the chamber to meet up with TEA-TEB. After combustion begins, kerosene is injected into the chamber, and the flow of the TEA-TEB igniter fuel is turned off. As the flow of oxygen and kerosene is increased, the engine's thrust increases.

SpaceX still dealt with ignition issues a decade later, after it began attempting to land Falcon 9 first stages. This entailed relighting engines not on the launchpad, but rather in the turbulent, hypersonic wind of atmospheric reentry. Among the lessons learned was including enough TEA-TEB to relight the Merlin engines multiple times after liftoff. Most notably, this problem occurred during the inaugural flight of the Falcon Heavy rocket in 2018, when there was not enough ignition fluid to light the outer two engines after several three-engine relights. The result of not relighting to make a landing burn? The center core missed its drone ship target by about the length of an American football field, slamming into the ocean at a speed of 300 m.p.h. The drone ship sustained minor damage. The rocket sleeps with the fishes.

In McGregor, after the ignition tests, the team was ready to turn on the rocket engine for real, although this process came with a series of staged tests as well. First, there would be a little bit of flame, then a partial power test, then a short, full-power test before, finally, a full-power, full-duration test. For the Merlin engine scheduled to power Flight Two, this full-duration firing test came in the last days of Dunn's internship, a perfect capstone on his hot summer in Central Texas.

Once they had prepped the engine, the team of engineers and technicians retreated a few hundred yards, into nearby fields. From there, close enough to be awed, but not close enough to be in real danger if something went wrong, they watched and waited for the test to begin.

Then, the engine blazed into life, an intoxicating mixture of fire and light, smoke and soot. "It was a monster," Dunn recalled of the Merlin firing in 2006. "I was just blown away by how loud and powerful it was. It felt insanely loud. It was perfect. I was absolutely hooked."

It had been quite a summer. Along the way, Dunn had benefited from the generosity of the older men he worked with. Some of these "veterans" weren't that much older than the twenty-four-year-old Dunn, but they had the experience of that first flight from Kwaj. Hollman and another propulsion engineer, Kevin Miller, took time to explain the Merlin's inner workings to Dunn. Between cigarettes, Eddie Thomas shared his tricks of the trade.

They also grumbled about the changing times as SpaceX matured from a rough-and-tumble start-up toward respectability. That year the company had won the $278 million contract from NASA, and gotten a new CEO in Jim Maser. Along with money and new leadership came new rules and procedures. All work on the rocket had to be documented, with fewer decisions made on the fly. There had been less of this during the run-up to Flight One, but now the company had evolved.

As they worked that summer, Dunn and his colleagues talked not only about the upcoming flight but also about their first launch just a few months earlier, in March. The older hands passed on their knowledge. What Dunn savored most of all were the stories, the tales of life at SpaceX when there were only a few dozen employees, when everybody knew everybody, and there were Friday afternoon ice cream runs. Those times were gone. He again worried that he had missed out on the company's formative years.

The early days *were* over. Musk and SpaceX were growing to meet the great demands of orbit. There were no longer a few dozen employees, but more than one hundred. And as SpaceX matured during the summer of 2006, so did Dunn. He'd always been a smart kid and a hard worker. Now he'd found a purpose, and that purpose would wring every ounce the young engineer had to give. In the end, Dunn had not missed the most important period in the company's history. He'd arrived right on time. Rather than being a mere bystander to the company's success, the kid from Tennessee would be pivotal to it.

After his summer internship ended, Dunn returned to Stanford, watching Flight Two from afar in March 2007. As that rocket had very nearly reached orbit, anticipation for the third flight ran high when Dunn hired on full-time at SpaceX in July 2007. He returned to Texas, and all through the late summer and fall of 2007, beneath the torrid Texas sunshine, SpaceX engineers tested the new, regeneratively cooled Merlin rocket engine at McGregor. This engine bore the name Merlin 1C. The original engine, powering the first two launches, had been the Merlin 1A. There had been a Merlin 1B, an ablative design of a Merlin engine with about 10 percent more thrust. But Mueller scrapped this project after work on a regenerative version proceeded surprisingly smoothly. Musk had been right. They should have gone with the regenerative design in the beginning.

The Merlin 1C engine performed like a champ on the test stand in Texas, firing for minutes at a time. No longer did they need to trash old thrust chambers after single, prolonged test-firings. The regeneratively cooled engine could go again. And so they continued to work through that winter, putting the new engine through its paces. After engineers had test-fired the actual flight engine through the equivalent of more than ten launches, SpaceX declared the new Merlin 1C ready to fly in February 2008. Soon, it would ship out for Flight Three.

This time, the engineers confidently believed they would reach orbit. And Zach Dunn planned to be right alongside them on Kwaj, doing his part to ensure success.

8

FLIGHT THREE

May 2008–August 2008

Zach Dunn was eager to get on with it. By the late spring of 2008, more than a year had passed since SpaceX's near miss on Flight Two. The company faced increasing financial pressure to get its third rocket out of the factory, onto the launchpad, and into orbit. No one hungered more for that moment than Dunn. Since begging Tom Mueller for an internship, Dunn had risen to every technical challenge put before him. So though he was mere months into his full-time employment at SpaceX, Mueller awarded Dunn with responsibility for the Flight Three rocket's propulsion system.

The first stage arrived on Kwaj in May. Dunn and a dozen other engineers and technicians flew in over Memorial Day weekend to prepare the stage for a static fire test in June. That static firing mostly went well, but as components of the rocket's second stage arrived on Omelek in June, a problem emerged with the engine's "skirt." Designed by Mueller, and managed by a former TRW engineer named Dean Ono, Kestrel powered the second stage during its flight in the vacuum of space. It does the in-space work of pushing satellites into orbit, and under Ono's watchful eye it never failed during tests. The skirt extension, a little more than four feet in diameter, expanded Kestrel's exhaust and increased its efficiency in a vacuum.

Shipped separately from the Kestrel engine itself, the skirt was one of the last pieces of hardware for Flight Three. Therefore, SpaceX wanted to get it to Kwaj quickly. The Army considered airlift a precious resource, so logistics officers had created a requisition priority system. When SpaceX first arrived on Kwaj in 2005, Tim Buzza and some of the other company leaders had cozied up to the Army's logistics liaison on the island. This relationship permitted SpaceX to designate most of its flight hardware with the code "999," typically reserved for critical war materiel. In its rush to ship hardware for the Flight One launch attempt in late December 2005, SpaceX's use of this highest priority shipping code filled the last Army logistics flight to Kwaj before Christmas. This bumped the delivery of Christmas trees, hams, and presents to locals until after the holiday. "The Army moms knew it was us," Buzza said. "They were fuming mad, to the point of calling us out in grocery stores. We learned to use the code a little less after that." Accordingly, in 2008, SpaceX took a different route to expedite the delivery of the skirt to Kwaj.

An employee flew with it commercially from Los Angeles to Honolulu, then caught a charter flight to Kwaj. When the skirt arrived, Dunn and the others dutifully made the trip to Kwaj from Los Angeles to assemble the second stage. On Omelek, they opened the package that contained the skirt. With the sun beginning to climb into the sky, an inspector looked over the skirt with a magnifying glass. Made from a niobium alloy, one of the hardest metals, SpaceX marketing materials at the time bragged that an orbital debris strike on the nozzle "would simply dent the metal, but have no meaningful effect on engine performance." The inspector, a quality assurance manager named Don Kennedy, did not find a dent in the nozzle, but rather a long, hairline crack in one of the primary welds. Dunn had been on the island for all of fifteen minutes. Yet the launch team could not assemble the second stage with a cracked Kestrel skirt. Back to Los Angeles they went.

"I don't recall ever being pissed off about that kind of thing," Dunn said. "I tried to consciously never be tough, or ride the folks who had these sorts of issues, because I always felt like it was a band of brothers, a band of sisters. We were all in this together."

Anger and frustration, he knew, would not serve any purpose. It could cloud one's mind during intense moments, whereas creativity or quick thinking could sometimes spark a solution. On Omelek, there was no help for the cracked skirt. But a few months later, this sanguinity would save the day for Dunn and SpaceX.

Later, the engineers pieced together what happened with the damaged skirt. The launch team was sensitive to the potential for damage to welds during transit, which was why they had sent an employee to chaperone its movement. They had also attached an environmental data recorder to the package. This data indicated the damage occurred on the ground in Hawaii, between flights. The SpaceX employee had taken a cab from the commercial airport in Honolulu to Hickam Air Force Base, where the transport plane awaited. A courier moved the skirt between airports.

"That was the only time that our watchdog wasn't with it," Buzza said. "The driver on that courier must have hit a two-foot pothole, or something, because we believe that's the event that cracked the skirt."

This just meant more delays for a flight that SpaceX already had a lot riding on. The company had not reached the financial breaking point yet, thanks to NASA's funding in 2006. But a company founded to profit from launching payloads into space had to, at some point, begin launching them safely into orbit. Following the near success on Flight Two, Musk declared that the Falcon 1 rocket had progressed from the development phase and into operational status. This was tantamount to saying the Falcon 1 rocket was ready for prime time. With the experimental Flight One, the company had flown a small satellite built by Air Force Academy students, and valued at less than $100,000. Flight Two launched a dummy spacecraft that approximated a payload but had no value. For Flight Three, SpaceX selected a trio of customers it could ill afford to disappoint.

The Air Force provided the primary payload, a 180-pound satellite named Trailblazer, to test new capabilities in orbit. NASA had two tiny satellites, one CubeSat to study the deployment of solar sails, and another to study yeast growth in space. Finally, SpaceX had its first truly commercial payload, the "Explorers" mission from Celestis, which sends cremated remains into space. On this flight the company gathered the remains of paying customers as well as several notable figures, including James Doohan, famed for his role as the engineer Scotty in the original *Star Trek* television series and films. After years of acting as a spaceflight pioneer in life, Scotty had a chance to reach the heavens in death. None of these were multi-million-dollar satellites, but in a single launch SpaceX was hosting payloads from its three most important customers—military space, civil space, and commercial space.

As always, Musk looked beyond the present mission to bigger and brighter things. During the spring and summer of 2008, he talked up plans for an upgraded version of the existing rocket, dubbed the Falcon 1e. And he spoke of rockets larger still. SpaceX engineers split their time between Flights Two and Three working the Falcon 1 program as well as pushing development of the Falcon 9 rocket with its nine Merlin engines. The company also began preliminary work on linking three of these Falcon 9 cores together to form the Falcon Heavy.

None of these ambitious plans would come to pass, however, should SpaceX fail to reach orbit with the relatively simple Falcon 1 rocket. Customers would scurry away. Other companies' rockets might cost substantially more, but at least their payloads didn't end up on the bottom of the ocean. NASA, too, would lose faith in the new space company. And Musk's funds, not unlike his patience, had limits.

"The crazy thing is that I originally budgeted for three attempts," Musk said. "And frankly, I thought that if we couldn't get this thing to orbit in three failures, we deserved to die. That was my going-in proposition."

By the time of Flight Three, the SpaceXers had grown accustomed to their visits out to the Kwajalein Atoll for launches. Over the course of three years they learned how to survive in the tropical environment, and even enjoy island life. Some of these lessons were hard-won, however.

Fairly early on during the Kwaj experience, Brian Bjelde missed the evening boat back to Kwajalein. It happened. He and a few others slept under the stars, passing a perfectly pleasant night. But the next morning, Bjelde lacked a change of clothes. So he grabbed a T-shirt from a package of Falcon 1 swag items that had shown up in Omelek. The vacuum-packed, white T-shirt may have been wrinkled, but at least it was clean, and it kept the sun off his back. Bjelde went through massive quantities of sunscreen every day—any piece of skin exposed to the tropical sun was covered. Throughout that day, as he slathered himself in it, Bjelde noticed the T-shirt's wrinkles straightening beneath the island's heat and humidity.

Late in the afternoon, he went to take a shower. "I took my shirt off, and I had the worst sunburn of my life through the t-shirt," he said. "I had a perfect sunburn. I think the end of my days will come through skin cancer because of that Kwajalein experience. The sunshine went straight through that cheap white T-shirt. I just didn't think to put sunscreen underneath it. Why would you?"

The heat and humidity punished those on Omelek in other ways. Bjelde had been a California kid, where it sometimes got hot, but rarely so humid. And he had never done this kind of physical labor before. Marines out at sea may be familiar with crotch rot, but Bjelde had never heard of it. "I don't claim to be thin or skinny, and if your thighs rub a little bit, and then you sweat, it leads to chafing," he said. "But the salty, humid environment just made things worse."

As he struggled to move around one day, walking bow-legged across the island, Bjelde asked the more experienced Chris Thompson what he should do about his painful ailment. Did he need penicillin, perhaps? Thompson, a former Marine, explained the trick of rubbing underarm deodorant between the legs. And Thompson offered another useful suggestion, switching from boxers to boxer briefs.

The few women on the island faced their own travails. In the early years, Anne Chinnery and Flo Li had little privacy and no running water. Using the island's toilet necessitated filling a bucket with sea water first, so the toilet could be flushed. Showers were even more rudimentary. Initially, the SpaceXers filled a trash bucket with water to wash their hands. When she got really hot and sweaty, at the end of a long day, Li said she would put on a bathing suit and dump the rain water over her head to rinse off.

As the Flight One campaign progressed in 2006, the small team moved on from a trash can to a camp shower. They collected rainwater in large, black bags, and then laid the bags out on the helipad to warm up during the daytime. One of these bags would be dragged onto a stand so that it hung over a folding metal chair, allowing for the luxury of a warm shower. For the benefit of Chinnery and Li, a shower curtain afforded a measure of privacy.

The engineers and technicians worked hard during the daytime, but as the sun neared the horizon, the crews would often take a break. They'd swim; a few people even skinny-dipped in the lagoon as a final escape from the heat.

Sometimes their efforts at rowdy fun would go awry. Omelek was small enough to walk across in a few minutes, but during later flight campaigns there was an old, raggedy golf cart for employee use. Dunn described it as "serious turd," hung together with bailing wire and bubblegum. At some point between Flights Two and Three, the brakes went bad on the cart, but when the launch team returned from Los Angeles no one realized it. At the end of one workday, as some of his friends caught the boat back to Kwaj, Dunn decided to see them off in style. He hopped onto the cart, parked near the trailers for overnighters, and put the pedal to the metal. Dunn thought it would be fun to zoom past his departing friends, honking and waving.

He had built up a good bit of speed as he neared the dock, and in preparation to salute the boat, Dunn decided he had better slow down. In a slapstick moment straight out of a cartoon, when Dunn pushed on the brake pedal it went down to the floor without any resistance. He succeeded in getting his comrades' attention, but for the wrong reasons, as he began screaming and careening toward a small rocky ledge. From there, Dunn faced a clear shot into the lagoon, likely flipping end over end as he went. He made a split-second decision and veered toward a palm tree instead.

"Instead of seeing me waving and honking and being a general goofball, they just saw me drive out at full speed, no explanation whatsoever," Dunn said of the group on the boat. "Then I crashed into a palm tree at top speed."

The impact threw Dunn over the cart's steering wheel, but he walked away from the accident. The employees on the boat laughed uproariously.

Some SpaceXers who stayed overnight fished the coral reefs surrounding Omelek, though they released anything they caught. Small organisms that grow on tropical coral reefs produce ciguatoxin, which accumulates in small fish, and in greater concentrations in larger fish on up the food chain. The Marshallese people developed an immunity to the toxin, but it causes severe food poisoning in outsiders. Every now and then the SpaceXers would hear a report of a visitor to Kwaj dying after eating a reef fish.

There were natural threats on land, too. Coconut crabs, which can grow up to three feet in length and are the largest arthropods in the world, lived on Omelek. Sometimes, they would be seen scurrying up a tree, and using strong pincers to knock a coconut to the ground. Then, back on the ground, the crab would crack open the coconut. "There was definitely no sleeping naked on the beach for us," said Jeff Richichi, a structures engineer.

By the time of Flight Three, the Omelek engineers and technicians had continued to refine their environment, especially with better food for those who slept on the island. In the double-wide's kitchen, they would take turns cooking meals that outclassed the fare in the Army cafeteria on Kwaj. In the mornings, they feasted on steaming plates of scrambled eggs. In the evenings, they mixed it up. Bulent Altan and a new launch engineer, Ricky Lim, did a lot of the cooking because they enjoyed it. It might be grilled steaks one night, or shrimp in paprika sauce the next. Altan's specialty was a Turkish goulash he loved to bake, mixing pasta with garlic and yogurt, smothered in a butter-and-tomato sauce. It proved a crowd pleaser on Omelek. There were other comforts, too. A refrigerated sea van had an endless supply of drinks, including beer for the evenings.

"Everything was fantastic luxury, compared to the first flight, so we loved it on Omelek," Altan said. "After the really, really crazy days everyone gathered around at dinnertime and really enjoyed just sitting down, and relaxing. We would always watch the same movies over and over again, like *Starship Troopers*. The most important thing is that the camaraderie was great."

The overnighters also built a wooden deck attached to the trailer. From there, they could survey some of the darkest skies on Earth. Oftentimes, clouds obscured their view. But when it was clear there were a million brilliant stars. Sometimes there were artificial stars, too. They looked like shooting stars but didn't fade out. Rather, they brightened.

Because these were intercontinental ballistic missiles being shot from the mainland United States toward Kwajalein Atoll.

It was a great irony: the imperative to fly fast pushed SpaceX from Vandenberg to Kwajalein, and once there, the employees had a grand view of missiles launched from Vandenberg. For the better part of half a century, the small atoll had served as ground zero for the development of intercontinental ballistic missiles and, later, President Ronald Reagan's "Strategic Defense Initiative." The Army's facilities on Kwajalein still serve a number of purposes, but the most enduring one is acting as a giant target range.

When the Air Force wants to test the accuracy of a Minuteman III missile, it will launch the three-stage, solid-fueled rocket from Vandenberg toward Kwaj. With its sophisticated radars, cameras, and other tracking equipment, the Reagan test site in Kwaj captures precise radar and optics data about the missile as it streaks through the atmosphere at about four miles per second. Often, the missiles targeted Illeginni Island, on the western side of the atoll. This means they passed almost directly over Omelek, on the eastern rim of the chain of islands. From there, SpaceXers spending the night on Omelek marveled as these missiles came in. Nearing Kwajalein, the missile's third stage would drop off, leaving only the missile bus, carrying the simulated warheads.

For some of the older hands on Kwaj, the prospect of incoming ballistic missiles rekindled distant memories of the Cold War. Seeing those rockets come in was beautiful, but also a little terrifying, knowing that if a real warhead was on board death would be imminent. "The simulated bombs would split off like little fireflies," Chinnery said. "It was very eerie to see those. It reminded me of growing up and being scared of nuclear annihilation."

Another advantage of spending the night in the double-wide trailer was skipping morning rush hour, and catching a few more winks. The big catamaran that carried Boeing employees from Kwaj to Meck Island,

and then dropped off SpaceX workers at Omelek, was reliable. But it got an early start, leaving the dock at 6:05 A.M. This meant that Buzza's team had to rise early if they wanted to eat breakfast before riding across Kwaj to the pier, to catch a ride.

"I never missed that boat, ever," Buzza said. "But sometimes, team-mates would. The Army is very punctual, and would not deviate. Except for one time, they did come back to the dock for Elon."

If an employee missed the big catamaran, there were other options. After SpaceX decided to shift its launch site to Kwaj, Musk purchased a fishing boat called the *Peregrine Falcon* and shipped it across the Pacific for employees to use. The *Peregrine*, as employees referred to it, had a big, open deck up front, and an area to sit behind the wheelhouse. A couple of people could also stand on the tower, to look for fish. About fifteen or twenty people could fit on board.

"Some days the ride was pretty smooth," Dunn said. "Other times we were going through eight or ten-foot waves, and just rocking and rolling. I loved to sit on the front of the boat on the way home, on days where it was rough, just getting splashed by waves."

At places where the ocean met the inner lagoon, waters were most furious. Meck Pass, a strip of water open to the ocean between Meck and Omelek, sometimes produced waves as high as fifteen feet. Employees riding on the tower might barely see over the crest of oncoming waves, but those on the lower deck could only stare into the oncoming wall of water.

The *Peregrine Falcon* wasn't really meant for this kind of daily, heavy use, spending hours each day in sometimes rough seas. It was always breaking down. Some time between Flight Two and Flight Three, Buzza heard through the Kwaj grapevine about a couple who had been sailing around the world. He asked them if they would be willing to operate and maintain the *Peregrine Falcon* for SpaceX. They would, and so thanks to "Salty Dog" and "Space Mom," the SpaceX boat became more reliable.

When an engineer or a technician missed the catamaran, and the SpaceX boat was down for repairs, Buzza faced a choice. Should he lose a day with that employee, or fly them out via helicopter? Patched-together Hueys offered one of the common modes of transportation around the atoll. All Buzza needed to do was call the flight line, and see which of the sandal-wearing pilots was available. The problem was, he began to recognize all the pilots. And invariably, if Buzza stayed up drinking at one of Kwaj's two bars, the pilots would be there, deep in their cups. He also noticed that as they island-hopped, the pilots never flew very high above the water. During one flight, Buzza asked why they didn't fly higher.

"I only fly as high as I can jump," the pilot replied.

As Flight Three approached, SpaceX was no longer a struggling band of a few dozen employees. Increasingly it resembled a bona fide rocket company. The large contract awarded by NASA in August 2006 allowed SpaceX, then spread across four buildings in El Segundo, to consolidate operations into a distinctive white factory in nearby Hawthorne. Its new address was 1 Rocket Road.

For years, Boeing assembled 747s inside the sprawling factory. But when Bob Reagan, the machinist hired by Musk in the company's early days, first saw the old Boeing building he came away unimpressed. "It was an ugly, ugly building," Reagan recalls. "I got elected to go and put the building together. It was the biggest nightmare I'd ever had."

SpaceX leased the property in May 2007, and Musk wanted the company moved in by the end of October. Over the summer, Reagan had to gut the building, add custom A.C. ducts, and more. He also had to configure the building with additional floors and office space that could eventually expand into 1 million square feet for the factory, mission control, cubicles, and dozens of meeting rooms.

After a legion of contractors met the seemingly impossible

deadline—the building was indeed ready for SpaceX's three hundred employees to move into by the end of October—Reagan initially felt jilted by the reward Musk offered. "He gave me a ten-thousand-share bonus, and I was so pissed off because I thought that was nothing," Reagan said, breaking into laughter. "I didn't know the stock was going to go up to $212 a share. I guess he took care of me."

Among Musk's talents as the leader of SpaceX was finding different ways to motivate his employees. Steve Davis said Musk would often visit his desk to ask detailed questions about his computer simulations for controlling the rocket in flight. And then they would make bets on some aspect of the rocket and its avionics system. Almost invariably, Musk would win. But ahead of one systems test in 2007, Davis said Musk raised the stakes. Davis bet twenty dollars he could complete some aspect of the test by a certain date. In return, Musk bet a frozen yogurt machine that Davis could not make the deadline.

"The second we had a bet like that, where there was a chance of getting a yogurt machine, there was a zero percent chance I was not getting that done," Davis said. "And if you go to SpaceX today in Hawthorne, you will see that he honored the bet, and we have a frozen yogurt machine sitting in the middle of the cafeteria, which still gives away free yogurt. So yeah, he's very good at motivating his people."

As SpaceX grew up, some of its earliest employees moved on. Phil Kassouf and Jeremy Hollman both left the company in November 2007. Kassouf left to pursue a master's degree. Hollman, meanwhile, had plotted his exit from the company for a while, and not simply because of his altercation with Musk after Flight One. Hollman still cherished the company's vision, as well as working with Mueller and the rest of the propulsion team. But he and his wife had married two years earlier, and now they wanted to start a family. Seeing the toll working at SpaceX took on Mueller's and Buzza's young families, Hollman felt as though he needed to move on to a less demanding job. As Mueller's chief deputy, however,

Hollman still played a critical role in rocket-engine development and testing, as well as assembling the rocket on Omelek.

Not wanting to let Mueller and the team down, Hollman sought out people at SpaceX he felt could carry on his work. One of his picks was Dunn. Hollman had worked with the eager graduate student during his Texas internship in the summer of 2006 and been impressed. When Dunn hired on at SpaceX a year later, Hollman tutored him in the finer points of rocket testing and assembly. He also groomed Dunn to take his place in the launch control room, to monitor the first stage's propulsion system both before and during the flight as data streamed back from the rocket.

When Hollman left, Dunn had been an employee for all of four months. All the same, Tom Mueller gave the kid in his midtwenties responsibility for the rocket's entire first-stage propulsion system. Dunn would oversee testing of the first stage in McGregor, and then travel to Kwaj during final assembly. "I can't say that I was replacing Jeremy because he was a legend," Dunn said. "I just went and tried to do as much of Jeremy's job as I could."

Responsibility brought a lot more pressure. After two failures, SpaceX had to get this one right. Not only did it have three customers on the flight, but more were watching and waiting before they booked a ride into space. And there was no shortage of competitors in the aerospace community just waiting to see another "commercial space" company fail, so the large launch companies could continue to collect lucrative government contracts without much competition.

"I didn't know what I didn't know," Dunn said of taking on this responsibility for the Falcon 1 propulsion system. "If something like that happened today, I would be super apprehensive. I would be very thoughtful about it. But at that time I was like, 'Let's just go get after it.' I just wanted to do the best I could. The way I approached it was to work my ass off."

And so he did. The first step toward flight involved finalizing the engine build. Only a few technicians and engineers did this work, beginning with screws, seals, O-rings, and transistors right down through full assembly. It entailed hustling around the factory to find parts, and taking pains to follow instructions. It took about a month to assemble a Merlin engine in those days, and this timeline involved fairly regular all-nighters.

"There was a certain amount of competitive, macho culture that was part of it, too," Dunn said. "Like I can fucking work as hard as anybody else. And I'm not going to be the one that has to go home first."

On August 3, 2008, Dunn took his position in SpaceX's small control room on Kwaj. As the Falcon 1 rocket counted down to launch, he watched screens that monitored the health of the Merlin 1C engine and the first-stage fuel tanks. Data rolled by, displaying various pressures, temperatures, and other variables. Though unable to sleep the night before due to anticipation, the prospect of an imminent launch kept Dunn in a state of hyperawareness throughout the morning. He'd come to Kwaj for the first time only a few months earlier and immediately loved the place. Hot and humid though Kwaj was, it couldn't match the heat of a Texas summer. And spending the night out on Omelek reminded him of camping back in Tennessee.

About a dozen engineers filled the small control room. As Dunn sat at his console, Hollman watched over his shoulder. His new employer, a Boston-based aerospace company, had agreed to loan Hollman back to SpaceX for this launch. Hollman didn't need to intervene, however, as Dunn had learned his new job well, closely following his handwritten notes and checklists. Like Hollman before him, he locked his eyes on data streams relaying information about the health of the Falcon 1 rocket's propulsion system.

Back in Hawthorne, confidence ran high. A festive atmosphere settled over staff members and their families as they gathered to watch the web-cast projected onto a large screen near the front of the new factory floor. Out back, Reagan had improvised a shot luge. He'd bought a four-foot long block of ice, engraved "SpaceX" into it, and cut a gulley down the middle of it. "Me and Gwynne took a tequila shot together," Reagan said.

It tasted like success. Everyone anticipated reaching orbit, and then throwing one hell of a party that night.

The countdown did not go smoothly. Buzza and his team had issues even before the launch window opened at 11 A.M. local time. The process of loading helium onto the rocket went more slowly than anticipated. This, in turn, caused the kerosene fuel already on board the vehicle to become too cold, not unlike the experience during Flight Two. The tank had to be emptied and the whole process restarted.

The launch window closed at 3:30 P.M. because this would allow the SpaceX team, in case of a scrub, enough time to detank and secure the rocket in daylight. As part of its range agreement with the Army, SpaceX had also run an analysis to prove that a launch during the allotted window would not collide with any known object in orbit. Outside of this window, there was a possibility, if an extraordinarily small one, that the Falcon 1 might strike an object already in space.

Everything seemed to be coming together for launch at the very end of the window, when Mother Nature intervened. At 3:20 P.M., during the terminal phase of the countdown, a thunderstorm moved directly over Omelek. The rocket could not safely launch into a storm, but the range weather forecaster predicted the system would blow through quickly. Knowing all else was ready but the weather, Buzza managed to negotiate a ten-minute extension of the launch window with the range commander. Soon, the storm did pass. At 3:34 P.M., Buzza gave the final go, and the Falcon 1 took flight. The white booster rose into the now-clearing sky, confidently soaring to what appeared to be a happy destiny.

The moment swept Dunn into something like an out-of-body

experience. Seated at his console, he lost track of time during the Falcon 1's ascent. Over the course of two minutes and forty seconds, the Merlin performed beautifully as the first stage climbed into space. For Dunn, it all passed in what seemed like less than a minute, a mere flash in time. And at the end of it, his first-stage propulsion system had done its job. Now the second stage would take over.

Then reality shook Dunn from his enchantment.

"When the anomaly occurred, I had my head down," Dunn said. "I was looking at data. And I could just hear this gasp. I looked up and by that point you could just see that things were not right. It took a little while to internalize. It was incredibly disappointing. The team was devastated around me." Some of his coworkers, in fact, were crying.

The gasp came from those watching the video monitor in the flight control room. A camera attached to the second stage, looking down, told the jarring story of what happened.

As the Merlin engine completed its burn, the rocket rose above the blue Pacific Ocean and white clouds far below. When the engine shut down, the spent first stage separated and began to fall away toward Earth. But then, before the first stage had dropped more than a few feet, it snapped back upward. To the horror of those watching, it thudded into the bottom of the second stage. In the kind of nightmare sequence that wakes a rocket scientist in a cold sweat, this collision sent the second stage tumbling out of control.

As the camera blinked out, it was clear the mission was doomed. The Falcon 1 rocket's first and second stages were plummeting back to the ground. Star Trek's Montgomery Scott had in fact breached the final frontier for the first time; not in life but in death, and not for eternity but momentarily. With this failure, it seemed like SpaceX's trek to the stars might be over, too.

In California, at SpaceX's headquarters, the celebratory mood quickly turned dreary. As usual, Mueller sat beside Musk during the launch, inside the mobile Command Van with access to the feeds from

Kwaj. Mueller felt good as his Merlin engine burned hot. But unlike the previous flight, the robust Kestrel engine would not get the chance to ignite, and boost its payload toward orbit. This felt crushing. The rocket had made it through the hardest part, the first stage, only to be tripped up again. To Mueller's eyes, from the video feed, it appeared as though the stage separation system must have malfunctioned, causing the collision. In the heat of the moment, he expressed this opinion to Thompson.

The structures engineer, who bore responsibility for this part of the rocket, reacted defensively to these recriminations, believing Mueller's conclusion premature. "That's bullshit," Thompson replied. "You've got to look at the data before you make that accusation."

Steve Davis was already looking at the data, and after an all-nighter, he figured out what happened first. As he pored over the video, frame by frame, Davis saw the actuators working. He confirmed there had been a full separation of the first and second stage. At another console he collected printouts of data sent back by the flight computer. And he noticed one curious bit of data, a non-zero acceleration of the first stage after separation. This, he realized, vindicated Thompson. Instead, Davis deduced the problem must have been caused by Mueller's new, regeneratively cooled engine.

Unlike its old ablative design, the Merlin 1C engine ran ambient-temperature kerosene fuel through channels in its chamber and nozzle. The propulsion team had not properly accounted for all of this fuel at the end of the first-stage burn. When the flight computer commanded the main engine to shut down, its software allowed for only a brief passage of time before signaling the first stage to separate from the second stage. But once ignited, a rocket engine will burn whatever fuel is available. And some of this residual fuel in the cooling channels combined with a small amount of oxygen in the chamber to produce a very slight, but catastrophic thrust.

"That one hurt almost as bad as Flight One, because it was so preventable," Mueller said.

Should Mueller and the propulsion team have caught the problem? Perhaps, but the residual fuel in the cooling channels really did produce just a tiny amount of thrust, and it lasted perhaps one second. At full power, the pressure inside the Merlin chamber reaches about 1,400 pounds per square inch. The transient thrust produced just after main-engine cutoff in Flight Three, by comparison, briefly produced a chamber pressure of 10 psi. This is less pressure than air exerts at sea level. So during the many engine tests in Texas, SpaceX had missed this transient thrust.

"You don't really even see it on the test stand, because the ambient air pressure is like 15 psi, and the rocket chamber pressure dropped to about 10 psi," Musk explained. "Later, when we looked back at it, if you looked super carefully you could see a tiny, tiny trail of thrust. But a rocket engine producing 10 psi in a 15 psi ambient atmosphere, basically you don't notice it. You don't see it in the data."

In the vacuum of space, with rocket hardware so close, even a miniscule thrust is enough to force a catastrophic collision between stages. The solution was to change a single number in the flight software. For Flight Four, all SpaceX had to do was add four seconds to the time in between main engine cutoff and stage separation. But this assumed there would be a Flight Four.

In the immediate aftermath of the failure, no one was making that assumption back on Kwaj. One of the engineers in the control room alongside Dunn, Bulent Altan, rode back to Macy's with his mind spinning like the tires on his bike. Had he and his wife moved from the Bay Area to Los Angeles and sacrificed so much for this? Would there be another attempt? Did Elon have any money left? Would there still be a SpaceX in a few days?

That night Altan and the rest of the SpaceXers on the island got really drunk, downing cans of Coors and Bud Light. They mourned what had happened. The first two flights had felt like a progression, from an early failure to a very late one just before reaching orbit. This third failure felt like a regression. And if the company was not moving forward, where was it going?

"Before that flight I think we all thought, 'This is the one, man. We got so far on Flight Two. We got this,'" Flo Li said. "That one was, I would say, the most heartbreaking flight for all of us. I think I felt the most devastated after that one just because we had gotten so far in Flight Two. I think we really thought that we had this in the bag for Flight Three, and then to fail in the manner which we did, that one was rough."

For every flight campaign, the launch team bought one-way tickets to Kwajalein. Due to the inevitable delays and slips in their schedule, the SpaceXers could never be sure when they would fly back to California. Only after the rocket launched would they buy return tickets. Following Flight Three, as the launch team drowned their sorrows, their humor turned dark. This time, the engineers and technicians mused, they might need to buy tickets back to Los Angeles on their own dime.

"Flight Three was devastating," Chinnery said. "Early on, Elon had said he would cover the first three flights. He wanted to give it a good college try. But how long would he stay in the game? Three failures is a lot of failures.

"Hardly anyone survives that in the aerospace world."

9

EIGHT WEEKS

August 2008–September 2008

Hans Koenigsmann felt like hell. The night after the Falcon 1 rocket's first and second stages collided, Koenigsmann spent long hours contemplating this disaster. Once again, he had to consider his role. Now chief engineer of launch, he carried some responsibility for failing to leave enough time between the Merlin engine's shutdown and stage separation. Like everyone else, he'd missed the threat of residual thrust.

It was bad luck, to be sure. But SpaceX had run through a lot of bad luck over the years, and bad luck only got you so far as an excuse. Maybe they just weren't all that good. Certainly, there could be no denying the company's dismal record. SpaceX had launched three times, with Koenigsmann playing an important part in all three missions. They struck out on all three. Musk had held up his end of the bargain, making good on his promise of providing seed funding, and supporting the company for three launch attempts. Now, with failure after failure, Koenigsmann worried that Musk might throw his dwindling resources and time into Tesla or another venture. He could hardly blame the entrepreneur.

The day after the third failure, Musk called a Falcon 1 staff meeting.

Dozens of employees crammed into the Von Braun conference room, located to the immediate left of the new factory's entrance. They sat at tables and stood along the walls of the trapezoidal shaped room. Musk took his place at the front, glass walls behind him, trying to find appropriate words for the moment. Koenigsmann, Buzza, and the launch team listened in from the control center on Kwajalein. Davis spoke first, walking those assembled through his preliminary findings on the Flight Three failure. It should be an easy fix, Davis said. Then it was Musk's turn. No one quite knew what the boss would say.

Musk felt as crestfallen as the rest of his employees. Worse, even. He had bet a lot on SpaceX, in time and money and emotional toil, with little return. Now, his personal fortune was running dry. He'd invested everything in SpaceX and Tesla. Beyond money, his personal life was falling apart. He and his first wife, Justine, had split that summer. They had been together since 2000, with Justine giving birth to six children. Their first son, Nevada, had died from sudden infant death syndrome at just ten weeks old. They had shared that horrible grief. Justine had also been with Musk at the beginning of SpaceX, by his side when the internet millionaire met a dirty and disheveled Tom Mueller for the first time in an industrial park. Six years later it all was unraveling. He'd tried to change the world, and the world resisted.

"At that time I had to allocate a lot of capital to Tesla and SolarCity, so I was out of money," Musk said. "We had three failures under our belt. So it's pretty hard to go raise money. The recession is starting to hit. The Tesla financing round that we tried to raise that summer had failed. I got divorced. I didn't even have a house. My ex-wife had the house. So it was a shitty summer."

Musk really had put all of his net worth into his rocket and electric car ventures, and in August 2008, he had almost nothing to show for it. His rocket company had produced a litany of failures. Tesla was equally cash-strapped, only just beginning to sell its first product, the Roadster, and an initial public offering was two years away.

As Musk looked around the conference room in early August, he saw a chance for salvation. He had a good team. He personally hired all of these people, judging them to be smart, innovative, and willing to give their all. He had driven them hard, so very hard. They had made mistakes. But they were dedicated, and had put their souls into SpaceX. So in this dark hour, Musk chose not to play the blame game. Certainly, he could dish out brutally honest feedback, crushing feelings without regard. Instead he rallied the team with an inspiring speech. As bad as Flight Three had gone, he wanted to give his people one final swing. Outside that room, in the factory, they had the parts for a final Falcon 1 rocket. Build it, he said. And then fly it.

What they did not have was much time.

"He surprised me," Koenigsmann said. "He collected everyone in the room and said we have another rocket, get your shit together, and go back to the island and launch it in six weeks."

After Musk's staff meeting, his employees realized they were playing for everything. If their final rocket launched safely into orbit, the company would have a chance to survive. Success would give Musk an answer for the company's growing legion of doubters. Shotwell, too, could stop trying to rationalize failure to potential customers and perhaps begin to ink new contracts. But if this rocket crashed and burned, well, everyone pretty much knew what that meant.

The period that followed would be the most memorable and arguably important period of the company's history, hardening its DNA and setting the stage for SpaceX to become the most transformative aerospace company in the world.

The fourth launch of the Falcon 1 rocket was originally earmarked for the Malaysian satellite that drove SpaceX to look for an equatorial site. Because the Malaysian government did not want to risk its satellite on an unproven rocket, Musk decided to fly his remaining

Falcon 1, then scattered in pieces in Hawthorne, as a demonstration mission.

It fell to Chris Thompson and the structures team to cobble together some manner of payload. As with everything else in the hurly burly aftermath of Flight Three, they had to hurry. Thompson worked with Jeff Richichi, who directed structures, and Ray Amador, who led dynamic modeling for SpaceX, to design something that would approximate a satellite. In less than a week the trio fashioned a chunk of aluminum into a 364-pound simulator to replicate the mass and shape of a commercial satellite. For a name, they strung together the first letters of their own last names to come up with R-A-T-Sat.

RatSat needed a logo before shipping off to Kwaj. Thompson always had a fondness for fast cars, and as a kid he remembered going to car shows and seeing lots of T-shirts emblazoned with cartoonish hot rods driven by a rat with bulging eyes. These "Rat Fink" designs were the product of a Californian artist named "Big Daddy" Ed Roth, who passed away a year before the founding of SpaceX. Thompson passed a Rat Fink cartoon along to one of SpaceX's logo designers, who produced a similarly stylized design of a cheeky-looking green rat wearing a red T-shirt with the initials RF. They plastered the logo on three of the satellite's six sides.

While Thompson finalized the mission's dummy payload, the rest of the Falcon 1 team worked to assemble and transport the first and second stages of the rocket from Hawthorne to Omelek Island as quickly as possible. Previously, the company had shipped the large first stage by sea. Although the second stage could fit inside a McDonnell Douglas DC-8 commonly used by cargo charter services, the first stage was too big. A truck would haul the Falcon 1 rocket's first stage down to the Port of Long Beach, where it would be loaded onto a container ship. Over the course of twenty-eight days, the cargo ship would offload other containers at ports in Hawaii and Guam before finally docking in Kwajalein. From there, another boat would deliver the rocket to Omelek.

But SpaceX did not have a month to wait for a circuitous barge trip. It would have to send the rocket using an oversized aircraft. Brian Bjelde, who had developed good contacts within the U.S. military during his work on the Falcon 1's flight termination system, dove into his Rolodex to try and arrange such transport. After reaching out to contacts at the Air Force, D.A.R.P.A., and other military offices, Bjelde received a callback. The Air Force had a C-17 aircraft available. "Somebody smiled on us," he said.

The Air Force alerted SpaceX that it would fly the C-17 aircraft into Los Angeles International Airport on September 3. In the month since Flight Three, the company's engineers and technicians had scrambled madly to complete assembly of the first stage. Dunn recalls that he and a good friend, Mike Sheehan, often slept at their desks for much of August, or powered through the overnight hours. "There wasn't a time in that month that one of us was not working on the rocket, and the vast majority of the time we both were," Dunn said. "It was always intense in those days, but this set the bar."

Dunn, Sheehan, and the rest of the Falcon 1 team managed to assemble the rocket before their Air Force transport arrived. This was an event they eagerly awaited. The company's new headquarters was only about five miles down Interstate 105 from the airport, and right along the C-17's flight path. The huge factory, with warrens and old catwalks, was mostly not yet built out, and its doors to the roof were open. Dunn and some of the others who had worked feverishly for a month climbed up to the top of the roof to watch the C-17 pass nearby, a sight to see above sprawling Los Angeles.

By then the C-17 had flown in active service for about a decade, playing a pivotal role in airlifts during U.S. military campaigns in Kosovo, as part of N.A.T.O.'s Operation Allied Force, and later in Operation Iraqi Freedom. The cavernous aircraft has a cargo bay that, at eighty-eight feet long and eighteen feet wide, could accommodate four big yellow school buses. When a U.S. president travels abroad, a C-17 follows behind Air Force One carrying the presidential limousine and Marine One helicopter. With a carrying capacity of 170,000 pounds, the aircraft could easily accommodate the Falcon 1 rocket's first stage. Fully loaded with fuel, the Falcon 1 had a mass of 60,000 pounds, but empty, it tipped the scales at only about 4,000.

Seeing their ride to Kwajalein amped up the launch team's anticipation. They transported the Falcon 1 rocket to LAX, meeting the Air Force crew at the back of the airport, near the ocean. "The Air Force people had been scrambled," Bjelde said of the flight crew. "They were all shaking their heads, saying this never happens. I don't know who the person was that said OK to the flight, but they may have saved SpaceX." The company would have to pay about $500,000 for the privilege, but the Falcon 1 had a fast ride.

The combination of a white rocket and the distinctive military transport aircraft made for quite a scene at LAX. Bulent Altan was among those helping to push the rocket onto the military transport as commercial jets passed on nearby runways. "I vividly remember a Virgin Australia plane going by," he said. "It was one of those big 777s, and at every one of the small little window holes there was a face looking in our direction. They were probably thinking that World War III had started, because there were these young kids loading a rocket into a C-17 in the middle of an international airport."

The SpaceX team did not just push the rocket on board and wave goodbye. To save time and the cost of commercial air flights, about twenty employees rode along, sitting in jump seats along the walls of the aircraft. After the Air Force finished buttoning down the payload, the C-17 took off from Los Angeles, and began climbing toward a cruising altitude above thirty thousand feet. Inside, the cargo compartment took on something of a party atmosphere. Wearing jeans and jackets, the SpaceXers relaxed and soaked up the moment. Steve Cameron, a propulsion technician, broke out an acoustic guitar. They were having the time of their lives.

No company vice presidents made the flight, so Chinnery, the operations manager, had responsibility for the Falcon 1 first stage until it reached Kwaj. After a while in the air, the pilots began to invite the engineers and technicians, one or two at a time, up the ladder leading to the flight deck for sweeping views high above the Pacific Ocean. And before too long, the islands of Hawaii appeared on the distant horizon. Preparing for the descent into Hickam Air Force Base just outside Honolulu, the SpaceXers returned to their jump seats and strapped in. They kicked back and rested their feet on the blue cradle holding the Falcon 1 rocket. For a blissful moment, it seemed like they just might pull this crazy stunt off.

And then they heard a loud, terrible, popping noise.

About halfway between Los Angeles and Honolulu, Altan had taken his turn to visit the cockpit. Altan is a great schmoozer, and after they learned of his background in avionics, the pilots were eager to show Altan all of the aircraft's displays and switch panels. He stayed there for hours, and as the C-17 approached the Air Force base in Hawaii, Altan belted into the observer's seat on the flight deck. When he heard the first loud pop, Altan thought the noise might have come from the airplane. But when it happened again a few seconds later, the pilots began speaking frantically to the loadmaster downstairs, over their headsets.

"I heard something about crumple, and rocket, and I realized it wasn't the plane but the rocket," Altan said. "That's when I just bolted downstairs."

In the main cargo bay below, it was bedlam. Upon reaching the floor, the first person Altan saw was his close friend and JLG companion, Flo Li. She was crying. Seeing Altan, she pointed to the first stage lying horizontal in the cargo bay. As Altan turned toward the rocket, his eyes swept the scene. A line of SpaceX engineers, faces white as ghosts, watched as their last chance to save the company imploded. The structure of the rocket caved in, one loud ping after another, as if some giant were slowly squeezing a beer can.

Initially, the engineers worried not for the fate of SpaceX, but their own safety. "The first thought I had was that this thing is going to implode and rebound," Chinnery said. "And it would kill all of us who were sitting next to the rocket in the airplane jump seats. So I hopped up, and told everybody to get to the front of the rocket."

As the launch team scurried to the front of the cargo compartment, Chinnery, Altan, Li, and a few others huddled to assess the problem. Quickly, they realized the rocket had imploded due to a pressure differential. The Falcon 1 rocket was designed for transport by truck and barge at sea level. The first stage had various breathers, vents, and ports, but most of these were closed for the flight to Kwajalein. In this transport mode, the large liquid oxygen fuel tank had only a small opening, a one-quarter-inch fuel line that went through a desiccant so that no moisture got into the rocket. Following takeoff, as the C-17 ascended, the ambient pressure in the cargo bay dropped. This posed no problem for the Falcon 1, as it was designed to be pressurized relative to its surroundings, like during a launch. During several hours of flight, the fuel tank's interior slowly equalized to the cruising altitude pressure. But as the aircraft began its descent toward Honolulu, there simply wasn't time for the pressures to equalize. For the LOX tank, it was like breathing through a straw.

SpaceX had prepared for this. Thompson had done the math, and so had Dunn, to determine how much vent area the first stage would need to maintain a stable pressure inside during the C-17 flight. The problem was that the manual provided by the Air Force had outdated information, and the descent and depressurization rates for the C-17 on the charter flight were significantly more aggressive than the figures provided in the manual. As the C-17 lost altitude, the LOX tank suffocated.

In their short meeting near the front of the rocket, Chinnery and the rest of the engineers soon realized what needed to be done to stop the rocket from imploding further. Either the pressure inside the airplane needed to drop quickly, or ambient air needed to be fed into the rocket. Better yet, both. After the brief huddle, Altan climbed back toward the cockpit.

"Hey, the rocket is crumpling, we have to go back up again," he shouted to the pilots.

Here, the pilots had a decision to make. They had a $200 million aircraft and two dozen lives to worry about. They were thinking it would be safer to simply open the plane's large rear door, and jettison the unstable rocket into the ocean below. And in fact, had no one from SpaceX been on board the aircraft, they would have done just that. But instead, they followed Altan's instruction. One of them replied "OK, boss," and the C-17 immediately began to climb.

Then a pilot told Altan, "By the way, we only have thirty minutes of fuel." This gave the C-17 time for one loop around Hickam Air Force Base before they lined up for a landing. Effectively, this meant the SpaceXers had about ten minutes before the aircraft would restart its descent.

Altan took the message downstairs. Climbing into the aircraft's main bay, he saw the SpaceX team pulling all manner of knives out of their pockets. "Everyone was already cutting into the white shrink wrap over the rocket," he said. "All of the other people from SpaceX had knives with them, which I thought was quite impressive on a flight."

No one had anticipated needing to open up the rocket in flight, so none of the SpaceX employees had brought any tools along for the ride beyond their knives. After a frantic search for something to work with, the loadmaster produced the C-17's meager tool chest, which contained a flat-head screwdriver and single crescent wrench. This, at least, allowed the technicians to open a couple of small lines. But to really equalize the pressure inside the rocket with that of the cargo bay, someone needed to open a large pressurization line leading into the liquid oxygen tank, which could only be accessed by climbing into the rocket's interstage.

The rocket continued to implode. All hell was breaking loose inside the aircraft. Amid the tumult and danger, Zach Dunn stepped forward to save his first stage. A few years earlier, Dunn had feared missing out on his chance to make a difference at SpaceX. Now he would climb inside a collapsing rocket, thousands of feet above the Pacific Ocean. He held a wrench—and the fate of SpaceX—in his hands.

The interstage lies between the first-stage and the second-stage fuel tanks. During launch, it protects the second-stage Kestrel rocket engine, and the outer structure falls away during stage separation. Before he entered, Dunn turned to his friend standing by his side, Mike Sheehan. If the rocket starts to blow, pull me out, he earnestly told his friend. To reach the pressurization port leading into the LOX tank, Dunn had to crawl all the way into the interstage. Darkness enveloped him as he moved deeper inside, along the wall. Only Sheehan's hands, holding on to his ankles, tethered Dunn to any semblance of safety. As he went, sharp components lining the exterior structure scraped his back. And all the while, the tank continued to pop and ping ominously.

Eventually, Dunn reached the pressurization line and managed to twist it open. To his great relief, he could hear air whooshing into the rocket. Above the noise, Dunn yelled to alert Sheehan he was ready to come out. Sheehan took this as a cry for help, yanking Dunn out of the interstage across the tangle of pressurization lines and valves. It hurt like hell, but Dunn emerged to find his efforts paying off.

The rocket hissed as it repressurized, and just in time, as the ten minutes allotted for dealing with the rocket had passed. As the C-17 began to descend again into Hickam Air Force Base, the SpaceX team could only return to their jump seats and catch their collective breaths, the stunned silence broken only by more bangs and pings, similar to those heard minutes earlier. Before their eyes, the metal stage began popping back into its cylindrical form. They could not know what it meant. The aluminum skin had not been intended to flex like this, as a rocket should never be exposed to higher external pressure.

In the month since Musk told his team to launch this rocket within six weeks, his team had scrambled to do just that, assembling the final pieces of the Falcon 1 rocket, and finding a means of rapidly transporting it to Kwaj. Everyone sitting in those jump seats shared Musk's passion for spaceflight. But in their haste to move the booster across the Pacific Ocean they had, at the very least, dented the rocket's fragile exterior. Moreover the rocket's internal structure, including first-stage slosh baffles, had probably been destroyed. Chinnery, Dunn, and the others were left to worry about what came next. "We all thought we were done," Chinnery said. "The tank had imploded. We were devastated." Even before the C-17 rolled to a stop on the tarmac in Hawaii, they began to think about shipping the booster back to their factory in Hawthorne, where they might have the tools to salvage it. And if it could not be fixed? They couldn't contemplate the implications of that.

Upon landing, the engineers walked off the plane. As soon as their mobile phones had signal, they began calling back to California to relay the bad news to the company's vice presidents. Chinnery's first call went to her boss, Buzza, the launch director. They had landed in Hawaii after dark, so the hour was well past midnight back in the States. As Buzza groggily answered his phone, he soon appreciated the gravity of the moment from Chinnery's trembling voice. Buzza also knew there was little to be done in the middle of the night in California, or by an exhausted crew in Hawaii. SpaceX had already paid for the flight all the way to Kwaj. Go to bed, he urged Chinnery. We'll talk in the morning. Trying to assuage her nerves, Buzza said that perhaps the damage could be fixed.

Meanwhile, Li called her boss, Thompson. She, too, was struggling emotionally with the collapse of the rocket. As the senior structures engineer on the flight, she had inspected the rocket and its fuel tanks after landing. From the outside, the Falcon 1 looked almost like nothing had happened. She asked Thompson if they should turn around and return to California. "You guys are going to keep going," he told Li. Try to get some rest, he added.

That was easier said than done. Due to the hastily improvised nature of the transport flight, the SpaceXers lacked sleeping arrangements at the base near Pearl Harbor. They had no ride to a hotel, or even a hotel to stay at. The military side of the airport lacked accommodations, so they crashed wherever they could. Some slept in chairs. Chinnery and a few others settled into a child's playscape, built near the main lobby of the airport. They bedded down on hard plastic slides, contorting their bodies to fit the curves. The C-17 air crew took pity on the SpaceX team and arranged for a few pizzas to be delivered. But even if the engineers and technicians had had the softest of beds and five-star room service, it's doubtful they would have slept well that night.

The refueled C-17 delivered the rocket to Kwaj the following day, and a flat-bottomed barge resembling a D-Day landing boat transported the first stage to Omelek. After the team rolled the Falcon 1 into the island's hangar, they performed a preliminary review of the rocket.

A tiny camera attached to a flexible tube, known as a borescope, was inserted through a sensor port into the first stage. About ten engineers and techs crowded around a tiny screen as the probe snaked around inside the LOX tank. "It was superhard to control the borescope probe, but all of sudden it flopped over and looked directly at a baffle that had been torn out of its bracket," Dunn said. "That was the moment when we knew for sure the rocket needed surgery and that we were screwed."

As the person in charge on the ground, it fell to Chinnery to come up with a recovery plan, assuming the Falcon 1 first stage could be salvaged at all. Following the company's formal procedures to methodically record operations and take a step-by-step approach to disassembling the rocket, she drew up a plan. Chinnery estimated it would require six weeks to take the first stage apart, inspect the damage, fix it, test it, and get back on track to launch. On Friday, September 5, she presented the timeline to Buzza, her immediate supervisor. Back in Hawthorne, he and Thompson shared it with Musk. "Elon saw that and went off the frickin' deep end," Thompson said. Six weeks was too long. SpaceX didn't have six weeks. Realistically, SpaceX did not even have a month before its funding ran out.

Thompson and Buzza retreated to a cubicle at the factory, calling Chinnery and a few of their direct reports. Inside the double-wide trailer on Omelek, Chinnery, Dunn, Sheehan, and a few other engineers gathered around a small table in a makeshift conference room, with a speaker phone in the middle. Chinnery opened the conversation to discuss her timeline, but before long Thompson cut her off. He felt like he needed to convey the gravity of the situation.

"You need to stop talking, and shut up, and listen to what I'm about to tell you," Thompson said. "You're not bringing that fucking rocket back. You're going to strip that fucking thing like a Chevy. And that rocket better be fucking disassembled by the time Buzza and I get there Monday morning."

There was dead silence in the trailer on Omelek as the words sunk in. They were going to have to fix the rocket right there, in the tropics. There was no time for quality control or meticulous records. They did not have six weeks. They had one. They were going to have to haul ass and hope for the best.

"There was a pause," Dunn remembers after Thompson spoke. "But it was really quick, with a transition to figuring this out. Engineers and technicians solve problems. We basically rallied up."

• • •

While the team on Omelek got to work, Thompson and Buzza prepared a rescue mission to help fix the rocket. At the factory in Hawthorne, they grabbed all the hardware that might be needed for repairs, such as baffles, and clips, and fasteners, and much more. Then they loaded Musk's Dassault Falcon 900 jet down with supplies on Saturday. Because there had been no time to ship TEA-TEB ignition fluid by boat to Kwaj, Buzza grabbed some of it as well. The container resembled a propane fuel tank, and as Buzza hefted it onto the jet, the pilot asked what was inside.

"Well, it's pyrophoric," Buzza replied. "That means when it's exposed to air it catches on fire."

"Will you fly with it in the seat next to you?" the pilot asked.

Buzza said he would. Asked what would happen if the TEA-TEB caught fire midflight, Buzza offered helpfully, "Well, you've got two choices: Go up really high and then depressurize the cabin so you don't have any air, or go really low so I can open the door and dump it out."

That was good enough for the pilot.

"We knew we needed TEA-TEB but we couldn't get it there any other way," Buzza explained. "This was some of the desperation that we were going through."

The Falcon 900 jet arrived safely in Kwaj on Monday evening, around nine, but Buzza and Thompson were not allowed to offload their hardware. Located to the west of the International Date Line, Kwaj is nearly a full day ahead of U.S. time. As a result, the Army facilities there took Mondays as their Sundays, and operated on low staffing levels. The skeleton crew checked the arriving flight in, but said its occupants would have to leave the airport until the next morning. As Buzza and Thompson departed the airport they worried about losing a half day, but luck was with them. Driving down the road outside the airport, they noticed an open gate near the jet. So the pair drove their truck through the gate and up to the plane, unloaded their cargo, and then proceeded straight down to the dock and the *Peregrine Falcon*. In the pitch-dark that night, they ferried repair parts out to Omelek.

On the island they found a hive of activity. Following the acrimonious telephone call three days earlier, Dunn, Ed Thomas, and other members of the propulsion team had returned to the hangar to remove the engine. To support the one-thousand-pound engine Thomas fashioned a makeshift platform from some wooden blocks. Working as fast as they could, Dunn, Sheehan, and others unhooked all of the fuel lines and other connectors linking the Merlin engine to the Falcon 1's first stage. For Dunn, it felt as though he were acting in a dramatic scene on a television hospital drama, with surgeons shouting out what they were doing and nurses rushing to provide tools. Off to one side, a couple of quality-assurance inspectors frantically struggled to record what was happening. In the span of a single hour they had stripped the rocket and put its engine on blocks.

Another team had worked to remove the raceway from the first stage. This is the assembly of conduits and cables running the length of the rocket. A third group began the process of disassembling the entire first stage. A day and a half later they had taken it entirely apart.

Working side by side, the engineers and technicians grew sweaty and dirty from turning wrenches all day. After sunset, the engineers polished their data-analysis tools, wrote procedures, and performed hardware reviews. By 10 P.M., they might finally knock off and grab a beer. Even back then, under the stars far from home, the SpaceXers understood they were different. Late at night, on the deck, they would joke about the rest of the aerospace industry. It was classic music, with good manners, bucolic countryside moments, and delicate discussions. SpaceX, by contrast, was hard rock and heavy metal. They were messy and loud, playing screaming guitars, and banging down the door. They felt this passion essential to surviving on the raw edge of the future and charging forward to build something great and new for the world.

By the time the vice presidents arrived on Omelek the engineers and technicians were ragged out and exhausted. But they had done the impossible.

"When Buzza and I showed up out there on Monday morning, that rocket was in fact stripped like a Chevy," Thompson said. "And so much so, that they took it to a whole new level, and they actually put the engine up on blocks. Which, by the way, was absolutely hilarious looking."

With Buzza and Thompson on the scene to supervise, and with replacement parts in hand, the launch team set about repairing the rocket. Broken slosh baffles were replaced, welds inspected, lines straightened. Within less than a week, they buttoned the first stage back up. Now, they had to test the integrity of the rebuilt stage. The refurbished LOX fuel tank still showed a few wrinkles, which raised concerns about buckling. Thompson figured that if they were lucky, the wrinkles might straighten out under higher pressures. And if they weren't lucky, well, it had been quite a ride.

Normally such a pressure test would be done by filling the tanks with an inert gas like nitrogen, which does not burn, and then slowly increasing the pressure inside. But the only commodities SpaceX had to hand on Omelek at the moment were liquid oxygen and kerosene fuel. This raised the stakes, for if the tank failed during pressurization with one of these propellants, it would do so catastrophically by blowing up.

"We knew full well that if anything failed, it was game over," Thompson said. "Believe me, it was a ballsy move. But I mean, that's the state that we were in. It was like, we have to make this work. There's no six weeks. This had to be done in days."

Risk be damned, the first stage wound up acing the pressure test. As a bonus, during pressurization some of the wrinkles in the LOX fuel tank did indeed smooth out. They had made the tank whole. Mere days after the fateful C-17 flight, the SpaceX engineers and technicians had fixed the Falcon 1 rocket first stage, tested it, and found it flightworthy.

"Of all the crazy things that we did over the years, and of all the amazing accomplishments in a short period of time, that one really stands out," Chinnery said. "I can't believe we disassembled an entire stage and reassembled an entire stage in the course of a week. I don't think I could have imagined that."

They had broken virtually every rule in aerospace to pull the first stage together, but because of these heroics on Omelek, SpaceX still had one last shot at survival. They pushed on hard throughout September, working late nights under starry skies, breaking only for grilled steaks or Turkish goulash. After the pressure tests, they bolted the second stage on to the first stage. Then the launch team rolled the entire rocket—the very last Falcon 1 hardware they had to hand—out to the launchpad. By the last week of the month they were as ready as they were ever going to be.

It was fly or die.

10

FLIGHT FOUR

September 29, 2008

Tim Buzza and Hans Koenigsmann sat in the kitchen of a small house they rented in Kwajalein, talking late into the evening. They were exhausted, but sleep eluded them as they discussed what else could possibly go wrong the next day.

Eight weeks earlier they had stood side by side in the SpaceX control room when Flight Three failed. After sharing that ordeal, the two friends spent fifty-six grueling days getting ready for this one final try. Now, as an air conditioner droned incessantly in the background, they pressed one another for things they might have missed in the rush to launch. Last time, a single line of code had derailed the rocket. On the eve of Flight Four, Buzza and Koenigsmann worried about what might trip them up this time. Finally, at midnight, they closed their laptops and decided to try and get some sleep.

Yet Buzza remained restless. He had put so much into this small company, having joined almost at the beginning. And for what? SpaceX would not be a real rocket company until it reached orbit. For all of the engine tests, static fires, and launches he had overseen—and they numbered into the hundreds now—SpaceX had not yet climbed to the top of the mountain. He left the house and got on his bicycle, riding to a place he often visited to clear his head. Just a few minutes away, North Point was the most barren spot on Kwaj, offering a clear view to the north, toward Omelek. He sat on the solitary park bench and let his mind wander.

Beneath the dark sky, he thought of his family. His wife and kids had sacrificed so much over the previous six years. Another failure would devastate them and dash his promises that success would make it all worthwhile. He also reflected on his launch team, who looked to him for leadership and confidence. These thoughts weighed heavily as the night deepened. Sagging into the bench, Buzza rolled his head back to look at the stars. He easily found the Southern Cross, a distinctive constellation of four bright stars visible in the lower latitudes.

"The brightest of the four stars twinkled a brilliant blue, and suddenly a calmness came over me," he said. "We were ready. I biked home and fell soundly asleep."

No such peace came to Zach Dunn. Outside the concrete walls of his spartan accommodations at the Kwaj Lodge, restless waves broke on the rocky beach. Inside, he tossed and turned, too. Jangled nerves and anticipation pressed on Dunn's awareness during the small hours as he contemplated the next day. His brief, meteoric career at SpaceX might screech to a halt within hours. Or it might take off toward boundless horizons. Whatever the fate of the Falcon 1 rocket held, he was eager to see it through.

Long before sunrise he rose from his rumpled sheets and dressed in the dimness. He walked from the Army hotel, a salt breeze upon him, and found his bicycle. For more than half an hour he turned the pedals over, riding toward the small SpaceX control center tacked onto a large

Cold War–era building. Even in darkness, the large Army defense facility cut an ominous profile against the night sky, looming ever higher above the palm trees as he rode closer. "It looks like a James Bond, *Goldeneye*, crazy-ass laser-beam facility," Dunn said.

As he entered the SpaceX office, Dunn passed through a small support room into the main control room. Inside, some five hours before the first opportunity to lift off, the SpaceX launch control team had begun gathering. The Army provided SpaceX with a generous launch window that day, from 11 A.M. to 4 P.M., local time. Even so, Buzza and Koenigsmann were already pushing the team to hit the beginning of it.

D ue to the time difference, it was midmorning in Southern California on Sunday, September 28, as Dunn pedaled across Kwajalein. Musk had not traveled to Kwajalein for the launch, opting to remain behind in California during the summer and early fall of 2008. He needed to be on hand for both SpaceX and Tesla as they struggled for survival, simultaneously managing operations and raising funds for rockets and electric cars.

It was a hell of a time to be running a single, cash-hungry start-up business, let alone two. The Great Recession, precipitated by a housing bubble and subprime mortgage crisis, technically began in the United States at the end of 2007 but began to swamp the wider economy in 2008. By the fall, overall economic activity in the country slumped as the U.S. gross domestic product fell by nearly 5 percent. Perhaps most critically for Musk, from 2008 to 2009, the amount of money U.S. venture capital funds raised dropped from $53.2 billion to $22.7 billion.

As the recession cast a pall over all speculative business efforts, Musk scrounged for funding to save both of his companies. While SpaceX whirled through the eight weeks between Flights Three and Four, Tesla's existence was no less precarious. The company had finally begun to deliver its first Roadsters and was in the process of unveiling the Model S car, but it also faced a cash crunch. Musk needed funding, and for that he needed results.

But on the morning of the twenty-eighth, mostly he needed to clear his mind. To kill time Musk and his brother, Kimbal, braved the weekend crowds to take their children to Disneyland in Anaheim. There, they rode Space Mountain, the famed, space-themed roller coaster. Was this an omen of some sort? "Everyone is superstitious in the launch business," Musk said. "Maybe it was a lucky move, I don't know. But after that I took them to Space Mountain a couple of more times before critical launches."

It takes about an hour to drive across Los Angeles from Anaheim to Hawthorne, so Musk had to speed along the 105 freeway to make it back to the company's headquarters by 4 P.M., when the launch window opened in California. Wearing jeans and a beige polo shirt, he rushed into the Command Van at SpaceX, took his customary seat to the right of Mueller. In front of him, a laptop provided data about the vehicle. Above him, on the trailer's walls, large monitors provided video of the Falcon 1 on the pad.

"I was stressed out of my mind," Musk said of the countdown. "Super tense."

Gwynne Shotwell was halfway around the world, quite literally, from Kwaj. She had traveled to Scotland at the end of September for the International Astronautical Congress, the world's biggest space conference. There, she had the unhappy task of briefing the Flight Three customers about the company's investigation into the failure, and their wayward payloads. In her time zone, the launch window opened at midnight.

Shotwell stayed up late that night after her husband, Robert, an engineer at NASA's Jet Propulsion Laboratory, went to bed in the hotel room. To not disturb him, Shotwell camped out in the bathroom, sitting on the toilet, and balancing her laptop on her knees. To hide the noise, she turned on the shower.

Shotwell spent much of her evening on the phone with Lauren Dreyer, a mechanical engineer who grew up in Central Texas and worked at the McGregor test site. Over the sound of the water, Shotwell and Dreyer discussed costs and rewrote sections of SpaceX's proposal for a lucrative contract with NASA. After SpaceX won the Commercial Orbital Transportation Services program award in 2006, the company had worked with NASA officials to develop the much larger Falcon 9 rocket and Dragon spacecraft. Already, many of the company's employees had moved over to this work from the Falcon 1 program. By late September, SpaceX was in the final throes of a competition to win a contract worth more than $1 billion to fly operational cargo missions to the International Space Station. It could be the answer to the company's cash needs.

But none of that would happen unless SpaceX could prove that it really did know how to fly into space. There was no way NASA would risk missions carrying hundreds of millions of dollars' worth of food, supplies, and science experiments on a company that could not reach orbit with a simple rocket.

When the clock approached midnight in Scotland, Shotwell broke off her discussion with Dreyer and opened a window on her laptop to watch the company's webcast, as well as a proprietary stream of data from SpaceX headquarters. The shower ran. Her husband slept. And Shotwell waited.

On Kwajalein the countdown proceeded with few hiccups. The Falcon 1 had almost no issues, and perhaps this stood to reason. This was the fourth time the launch team had done this, after all. SpaceX also had more people than ever on the island, about three dozen, supporting the flight. They should be getting better at countdowns with experience.

Shortly before the rocket entered the terminal countdown phase, at about ten minutes before T−0, Buzza shared a few final thoughts with his team. They all knew the stakes, he said. Focus on your last-minute tasks. Tighten your seat belts. Then he told the team they reminded him of the early NASA flight controllers who had guided humans safely to and from the Moon. They were also mostly in their twenties inside Mission Control during the 1960s, too.

"I was the grandfather," Buzza said. "I was about forty years old, and everybody else was clearly under thirty. The entire room was under thirty except for probably Hans Koenigsmann, myself, and one or two others."

And then it was time. At 11:15 A.M., just a quarter hour after the launch window opened, the Falcon 1 rocket reached the end of its count-down. Immediately after T−0, humans lose all control over a rocket. The only interface, an essential one, rests with the range operator, in this case the Army. An officer sitting at a console can send a destruct signal to the booster if it veers off course. But beyond that, the rocket's computer controls its flight. "After the launch you can't do anything," Koenigsmann said. "You're just watching it. We're sitting on the console, but it has no bearing on the outcome."

So they watched. The white rocket, with a stark black interstage, stood on the launchpad venting oxygen into the tropical breeze. Palms near the rocket swayed in the wind. Then smoke and fire burst forth as the Falcon 1 rocket rose from the launchpad. About twenty seconds into the flight, the video feed switched to an onboard camera, looking down at tiny Omelek Island, a receding speck in an endless expanse of azure ocean.

After a minute, with the rocket already more than a dozen miles into the sky, calls of "nominal" spread through the control room. The first stage and the Merlin engine, as it had done on the previous two flights, burned as intended toward space. Soon it came to the pivotal moment of stage separation. After about two minutes and forty seconds, the Merlin engine shut off. And then the onboard computer counted 1, 2, 3, 4, 5, and 6 seconds before the first and second stages detached from one another. This added time allowed for a safe separation. As he watched the monitor, Dunn saw the first stage fall away from the rocket, its job done.

"It was just an absolutely spectacular moment," Dunn said.

But it was not the final moment. SpaceX had seen the Kestrel engine burn before. And after several minutes, back in 2007, they had then watched the upper stage spin out of control.

In California, Mueller followed along at Musk's side. "The early ones were always tense," the chief of propulsion said of launch day. "I mean like butterflies, like sick. But tired, too, because I couldn't sleep."

If the worst happened, Mueller knew he probably would share the blame. "You know, it's usually propulsion that fucks up. Like it's known that 40 percent of all rocket failures are due to propulsion. Almost half. So as soon as separation happened, we thought we had it, right?"

As the Kestrel engine lit there were hugs all around in the Command Van. They were jubilant, but also mindful of the Flight Two failure. After a few moments of celebration, Mueller, Musk, and the others went back to their monitors. Kestrel had minutes to burn before the rocket reached orbit.

This was the moment when Steve Davis really began to sweat. He stood in the van behind Musk, thinking about slosh. Because the second stage had not ignited and flown during Flight Three, that particular risk had not been retired. Maybe the baffles they added would not address some underlying problem. "I was crazy nervous watching at that point," Davis said. "I couldn't sleep normally for the year and a half between Flight Two and Flight Four."

. . .

Jeremy Hollman had left SpaceX nearly a full year earlier. Although he returned to Kwaj during Flight Three to help Dunn and some of the younger engineers, the launch team felt confident enough by Flight Four to proceed without the veteran's guidance. And so Hollman, with so much time, toil, and travel invested in the Merlin engine, followed the launch from his new home in Quincy, Massachusetts.

Hollman had never watched a launch with his wife before, as he had always been at the Pacific launch site. It was nice to share the moment, but difficult to experience it as a bystander.

"Like the rest of the world, I watched it on the webcast," he said.

As the Falcon 1 rocket rose, and the first stage separated, Hollman studied the grainy webcast for possible signs of something going wrong. He couldn't see anything.

At around two minutes into the flight, the payload fairing split away from the top of the rocket. The camera on the Falcon 1's second stage captured an arresting view of the two fairing halves falling back to Earth. And still, the Kestrel engine continued to burn. Named by Mueller after the smallest of falcons, and weighing only about one hundred pounds, it glowed red as it consumed liquid oxygen and kerosene to push the upper stage toward a stable orbit around the Earth. Then, nine and a half minutes after launch, the Kestrel engine shut down.

RatSat had reached orbit.

"When Kestrel shut down, the place just exploded," Dunn said of the Kwaj control room. "We went absolutely wild. We were all jumping around. Hugging each other. Screaming. It was a righteous celebration."

Anne Chinnery worked as a vehicle controller for the launch, sitting at a console and sending commands to the rocket while it remained on the pad. She had initiated the final sequence to launch the rocket.

"By Flight Four we had gotten off the ground several times," she said. "Launch is just thrilling, always. But it wasn't a new thing any more. It got off the ground. And we're thinking, oh my gosh, oh my gosh, is it going to work this time? And then it worked."

Brian Bjelde had come to Kwajalein for the fourth flight as the mission manager, so he held court in the outer room, watching with officials from the Federal Aviation Administration and potential customers. Bjelde knew he might soon have to look for another job if this rocket failed. As the rocket flew, he had time to contemplate all that had happened in his five years at SpaceX, and more recent events on Omelek. It had all been such a whirlwind. Nerves and passion overcame him as the Falcon soared.

"It was awesome. I get emotional thinking about it," he said, his voice cracking. "I'm sorry. I do. It was chilling. It was huge. It was just validation."

The flight did not end once RatSat reached orbit. The launch team continued to watch and wonder about their bird for half an hour. Another Kestrel engine burn was planned for forty-five minutes into the launch, commonly needed to position a satellite for its final orbital insertion. A tracking station on Ascension Island in the Atlantic picked up the Falcon's signal, and the second burn went smoothly as well.

The jubilation had become unrestrained in the Command Van back in Hawthorne. They hugged more. They cheered. They had succeeded.

"It was just the dream making orbit," Mueller said. "There had been just so many simulations, so many thoughts about it. So, so much sweat going into it. So, you know, like, it was just our whole life. I mean we're working our asses off just to get to that point. It was like, I mean, what a relief."

A few minutes after the Falcon 1 made orbit, Musk walked out onto the factory floor, where more than a hundred employees had watched the launch. David Giger stood among them. He'd served as the mission manager for Flight One, and now he led propulsion development for the Dragon spacecraft. Thinking back to his own days on Kwaj, Flight Four felt pretty surreal with the fate of the company in the hands of a few dozen engineers far away. Giger worried about letting family and friends down if the Falcon 1 failed. He worried about letting the country down in a sense. If SpaceX went under, it would take a lot of the nascent aspirations of the new space movement with it.

Like during Flight Three, SpaceX had urged employees to invite their families to the factory for the launch. As some restaurants do with a kid's menu and crayons, the company passed out a one-sheet handout for children. They could do a word search for terms like *Falcon* and *Kwajalein*, play tic-tac-toe, or color in a mission patch. This patch, for the first time, included two green four-leaf clovers. For every launch since—remember, rocket scientists are a superstitious bunch—the mission patch design has included at least one four-leaf clover. As the Falcon 1 launched, Giger recalls a sense of quiet expectation on the factory floor. "It was a bit more reserved until they made it to orbit," he said. "But then it was craziness."

The crowd quieted down when Musk entered the cafeteria area. He gave a short, three-minute speech that was quintessential Elon. This was, he said, "one of the greatest days of my life."

But as ever, there was more work to do. Musk's mind was on Mars that afternoon. "This is just the first step of many."

Finally, Musk said they were going to have "a really great party tonight." And they would.

• • •

The party had already begun down the road at the Purple Orchid, one of the company's favorite drinking spots. The tropical bar and tiki lounge was located in El Segundo, less than a mile from SpaceX's original headquarters at 1310 East Grand. Some employees watched the flight from there, where the bar broadcast the webcast. Phil Kassouf, who had left SpaceX a year earlier for graduate school, had joined the watch party.

The SpaceX veterans had developed a simple launch tradition. If they succeeded, they were going to drink. And if they failed, they were going to drink. Until Sunday, September 28, 2008, they had never drunk to a successful launch.

Like an old ballplayer watching a son or daughter take the mound for the first time, it was difficult for Kassouf to stand still as others turned knobs and pushed buttons on Kwaj. "I actually couldn't watch," Kassouf recalled. "It was hard to watch." After the Flight Three failure, Kassouf worried the Falcon 1 might be snakebitten. What would it be this time? An errant semicolon in the flight software? Kassouf intimately knew some of the hardware riding along. Near the top of the Falcon 1, in the avionics bay where the flight computer commanded the vehicle, he'd built some of the printed circuit boards routing signals throughout the booster. And fly they did.

RatSat and the second stage it is bolted to have kept going in space ever since, at an average orbital height of nearly 400 miles. As of early 2020, RatSat had descended only to about 385 miles, and will likely remain in orbit for another fifty to one hundred years, estimates Jonathan McDowell, a Harvard University astrophysicist who tracks satellite launches.

"How cool is that?" Kassouf asked. "You've got boxes that you built, that you poured blood, sweat, and tears into, and they're in orbit for a century. That's when that surreal feeling kicks in, you know?"

As late afternoon turned into evening, and evening turned into night, the California parties swelled. Some employees had gone to the Tavern on Main, some to the Purple Orchid, which was where the party went latest that night. The drinks were put on corporate cards. After Musk finished a news conference and interviews, he made an appearance at both of the boisterous parties. When he walked through the door at each one, the assembled employees went nuts. Through his singular leadership, they had done a great thing. And they loved him for it.

"It's going! It's going!"

Robert Shotwell awoke to these words. No shower on Earth, let alone in a Scottish hotel bathroom, could have masked his wife's joyous shouting as the Falcon 1 climbed into space.

After the rocket reached orbit, Shotwell ran out of her hotel room and down the hallway to find the rest of her Falcon 1 colleagues staying there for the conference. Dressed in a pajama top and yoga pants, she screamed and hollered as she went. When Jonathan Hofeller, who managed sales to the Middle East and Asia at the time, opened the door to his room, she gave him a kiss and a hug. Shotwell, Hofeller, and a couple of other SpaceX employees went down to the hotel bar. It was now past midnight, and the bar had closed. They cajoled the hotel to reopen the bar, and the small group ordered champagne. Warm though the bubbly was, they drank it anyway. "It was terrible," Shotwell said of the champagne. "But that night in Glasgow was awesome."

The next day, on the opening day of the conference, she had been due to brief customers on the Flight Three failure. "I was supposed to be briefing the sad story of the Falcon 1, Flight Three to these unhappy customers," she recalled. "I said, 'Fuck it, I'm going to talk about Flight Four.' So I did a little bit on Flight Three and then talked about Flight Four."

Just as Musk found the lucky charm of taking his children to Space Mountain, Shotwell, too, developed a space superstition after Flight Four, what she calls her "launch juju." On subsequent launch days since that first, sweet success, she has plastered the inside of her shoes, almost invariably high heels, with yellow sticky notes upon which she has written Scotland. "This way I am in Scotland for every launch," she said.

Jeremy Hollman could not sleep for the rest of the night. Shortly after the launch, he called Mueller, who was still in the Command Van. After a few minutes, Mueller passed the phone around so that his former lieutenant could talk to other members of the propulsion team. Many of them remain good friends even today. Even so, Hollman could not help but feel somewhat detached as he listened to the merriment in the background of the phone call. Afterward, as night fell over the East Coast, Hollman was left alone with his thoughts.

"I had mixed emotions of guilt for not being there with the team, a hint of jealousy for the same reason, and a lot of joy for getting to experience a launch with my wife," Hollman said. "It really was the first moment it hit me that I wasn't a part of SpaceX any more. I was both sad and OK with it in the end."

He need not have felt guilty. Before leaving to start a family, Hollman wanted to put key people in place at the company to carry on his work. He'd done that, and the success of Flight Four secured the legacy of the Merlin rocket engine. And there was one more wonderful surprise waiting for Hollman. The couple didn't know it yet, but they had conceived a child just a few weeks earlier. Hollman had left SpaceX to start a family with Jenny, and they had done just that.

• • •

Many of Hollman's best friends remained at their consoles on Kwaj after the launch webcast ended. First they waited to see the Kestrel engine relight, and then they watched to see when the second-stage battery would die. It held out just long enough for the ground control station at Kwajalein to register the second stage and RatSat as the pair flew over the launch site.

"It was amazing to see something come back that you had just launched an hour and thirty minutes before," Koenigsmann said. "That was a pretty good illustration of what an orbit around the Earth means."

And then, there was nothing left to do but celebrate. Buzza, Koenigsmann, and the other engineers locked down the small control room, and then most of the launch team headed toward the dock. They rode their bikes madly beneath the tropical sun, out of their minds with joy. And as they pedaled, they called out a single word. *Orbit.*

They reached Kwajalein's dock as the *Peregrine Falcon* carrying the mission's fallback crew pulled in. Early that morning, this handful of employees had turned valves on Omelek, and then retreated to the safety of Meck Island to observe the launch from a bunker. Some of them had snuck outside to watch it in person, with their own eyes.

When the boat docked, the group's jubilation doubled. "Everyone just started chanting 'Orbit,' over and over again," Dunn said. "It was amazing to see those guys, and have the team reunited. And absolutely, the party continued from there. The whole island knew SpaceX, and what we were about. They knew we had had a hard time before that, and they were pulling for us. So I think basically the entire island partied as hard as we could that night."

They ended up at Veteran's Hall, which housed one of the two bars on the island. As Chinnery drank with her friends and coworkers, she could not stop thinking about how hard they had all worked to reach this moment. "What kept going through my mind is that we had just made history," she said.

The team had a mixture of feelings as they drank their way through cases of beer at Vet's Hall. Relief. Excitement. Awe. Overlaying all of this was the sheer exhaustion of working nonstop between Flights Three and Four. Somehow, in the darkest of moments, in the most distant of tropical outposts, with the final chance before them, they had pulled together. They all knew that, very easily, they could have been drinking away their sorrows that night, making final goodbyes before they dispersed to other rocket companies, into academia, or elsewhere. Instead, they toasted to their shared experiences and a bright future.

"That, maybe more than anything else, is what I've loved about SpaceX," Dunn said. "Knowing that all those folks around you and next to you, in the control room or wherever, they went through it, too. They lived through being pushed as hard as they possibly could, or further, and gave it their damnedest."

Most of them got roaring drunk that night. With no privately owned automobiles on Kwajalein, civilians cannot drink and drive. However, military police nonetheless patrol the island in golf carts, giving tickets to inebriated bike riders. Very late that night the SpaceX engineers sent two older merrymakers off from Vet's Hall, wobbling on bicycles down the road, away from the lagoon. These decoys—Salty Dog and Space Mom— had not been drinking. The rest of the SpaceX team waited for the police to take the bait and then, as quietly as they could, giggling in hushed tones, rolled down to the lagoon.

There, most of them stripped. The warm water beckoned.

11

ALWAYS GO TO ELEVEN

September 2008–May 2020

After sinking six years and $100 million into SpaceX, Elon Musk finally had a real rocket. Only a handful of countries had ever built a liquid-fueled booster and launched it into orbit before. With the success of Flight Four, the scrappy company from southern California joined an exclusive club whose only members were nations and state-backed rocket companies. In interviews immediately after the launch, Musk called the Falcon 1's brilliant flight "the culmination of a dream." And yet as he celebrated with his employees at the factory, and reveled in their cheers at the Purple Orchid, Musk lived anything but a dream.

Inside, it felt more like a nightmare.

"The thing is, I think my cortisol levels were clinically high, so I wasn't actually feeling celebratory," he said. "There was no jubilation or anything. I was just too stressed. It's like the patient survived. Getting to orbit was just like, OK, we're not going to die now. At least we'll live a little bit longer. That's what that launch meant. I just felt relief."

His employees did not fully realize the desperation of the situation, and Musk did not want to ruin their moment. But he had good reason to worry. Although the Falcon 1 success injected much-needed validation into the SpaceX brand, the company had no revenues it could quickly tap into. After watching the demoralizing failure of its first three rockets, no potential clients were calling Gwynne Shotwell. In fact, the company had just a single customer left for the Falcon 1: the Malaysians. Even if Shotwell's phone started ringing in the fall of 2008 after SpaceX finally reached orbit, those missions could not fly soon. SpaceX had no more Falcon 1 rockets in the factory and would not get paid until it put customers' satellites into space.

Meanwhile, cash burned away on all manner of fixed costs. The company had rent to pay for its extensive facilities and ongoing equipment leases on the tools needed to build engines and rockets. When Flight Four launched the company payroll exceeded five hundred people, and they expected paychecks, medical care, and other benefits. An early August investment of $20 million by the Founders Fund, a technology venture capital firm, helped. But after the fourth flight, SpaceX's finances remained dire.

"I had fantastic employees, and it was my job to make sure they could get paid," Shotwell said. Orbit or not, by that fall the money was all but gone. "I looked out six or eight weeks in advance, and knew we weren't going to have enough money to make payroll."

Musk's travails during this period are well cataloged in his biography, written by Ashlee Vance and published in 2015. He dealt with unrelentingly negative headlines in the summer and fall of that year, with the third failure of the Falcon 1, the creation of web sites like "Tesla Death Watch," and his ex-wife, Justine, dragging her former husband publicly in the press. His new girlfriend, an English actress named Talulah Riley, said Musk looked like "death itself" and described him waking from nightmares, screaming and in physical pain. She worried he would break under the stress, or perhaps have a heart attack and die.

Even as SpaceX achieved success, both of Musk's major companies spiraled toward bankruptcy. That fall he had about $30 million cash left. Friends urged Musk to save SpaceX or Tesla, warning that he could not support both. He agonized over the decision. "It was like having two children," Musk said. "I could not bring myself to let one of the companies die." In Musk's worldview, he could not let either venture go. Tesla was needed to save Earth from climate change, helping to break humans from their fossil fuel addiction. And SpaceX would offer a backup plan by making humanity a multiplanetary species. He split his money between the two companies.

Amid these bleak financial times, SpaceX had one final card to play. In 2006, NASA had come through with critical funding for the company after the failure of the first Falcon 1, betting that SpaceX would eventually figure out how to reach orbit. Even as the countdown clock for the fourth mission ticked down, Shotwell had worked to finalize the company's submission to NASA for the operational phase of the program, known as Commercial Resupply Services, or CRS. Through this contract, the space agency asked for help keeping astronauts on board the International Space Station fed and clothed. The contract would pay SpaceX to build Falcon 9 rockets and Dragon spacecraft to fly food, water, supplies, and science experiments to the International Space Station. This was the one pot of money that could buy SpaceX financial stability.

"It was not like we had a bunch of customers lined up," Musk said. "We had the Malaysians, and there was just not a good runway of stuff to do after that. Without the CRS contract, we would have gone down as that company that made it to orbit, and then died."

After winning the COTS contract from NASA in 2006, SpaceX put the money to good use. The company's workforce swelled to tackle these ambitious new projects. No private company had ever launched a spacecraft and then returned it to Earth, as SpaceX intended to do with

Dragon missions carrying tons of cargo to NASA's orbiting laboratory. Even as one team sought to finally bring the Falcon 1 to orbit, other groups began designing the Cargo Dragon spacecraft, and the much larger Falcon 9 rocket to launch it. As early as 2007, a majority of the company was working on the new programs.

Musk always intended to build a larger rocket, but he originally envisioned going from one to five engines, with a vehicle called the Falcon 5. This, he believed, was just large enough to nudge a small capsule into space. The COTS contract allowed Musk to think bigger. The space agency made clear that it wanted the capability to send multiple tons of food, supplies, and other equipment into space on each mission. Because NASA asked for a larger spacecraft from SpaceX, that meant it needed a beefier rocket to lift it into orbit. This led to the Falcon 9.

For years, the propulsion team struggled to reliably light a single Merlin at liftoff. Now they had nine to worry about. Mueller and his team also had to learn how to safely cluster these engines together—if there's a problem with one engine during flight, how close can others safely be located without catching on fire as well? Given these challenges, Musk initially thought it might be simpler to just develop a larger, more powerful version of the existing engine, called the Merlin 2. This would avoid the complications of clustering and controlling so many engines on a single rocket. But SpaceX could not afford the time or money such a project would have entailed, so the plan changed to multiple engines. "We knew that was going to be rough," Mueller said of a rocket with nine engines. "But we really had no choice."

By June 2007, SpaceX finished building its first fuel tank for the Falcon 9 rocket and shipped it to the McGregor test site. There, for the first time, engineers put the massive tripod Andy Beal had built nearly a decade earlier into service. In November of that year, they attached a single engine and test-fired it. By the following March they had fired three engines. To their surprise, igniting multiple engines simultaneously came off reasonably smoothly. All of the effort Mueller, Hollman, Buzza, and others invested into making the Merlin 1A and Merlin 1C rocket engines run efficiently for the Falcon 1 paid dividends when they bolted the Merlins onto the Falcon 9 rocket. Most of the kinks had been ironed out. So although there were more of them to manage, the Merlin 1C engines were known quantities.

During the summer of 2008, as Buzza and the Falcon 1 launch team prepared for the third and fourth flights on Omelek, a separate team in McGregor fired up a Falcon 9 rocket, with all nine engines, for the first time. These first tests lasted only for a few seconds. The bigger test would come with a full-duration firing that fall, in November. With the Falcon 9 rocket securely clamped to the tripod, its Merlin engines fired for 178 seconds to simulate an actual first-stage mission to space. Buzza watched from inside the blockhouse. Two months earlier, he'd sent a Falcon 1 into orbit for the first time, and now a rocket ten times more powerful shook the tripod and blazed brilliantly into the Texas night. "At the time, it was probably the most powerful thing I had ever seen," Buzza said.

The Falcon 9 rocket had arrived.

When he was not working on the Falcon 1 and Falcon 9 rockets, Mueller also led the design of new thrusters for the Dragon spacecraft. He tapped David Giger, a new engineer on his staff, to lead propulsion on this capsule. Dragon had to be versatile, autonomously controlling its flight through space in order to rendezvous with the International Space Station, and safely splash down in the Pacific Ocean. Giger and a small team of engineers started from scratch in 2006 when they began thinking about what a modern space capsule should look like. "Dragon was sort of this side project while most of the company worked on Falcon 1," Giger said. "I remember having Saturday meetings with Elon and a very small team of maybe five people. We were kind of just hammering out a few of the high-level concepts of Dragon."

SpaceX had some help from NASA. A small team of NASA engineers supported SpaceX and the other COTS contract winner, Orbital Sciences, when it came to reviewing spacecraft designs and highlighting potential issues. For a time, these NASA officials assumed they were working on a backup plan to supply the space station. The program was a low priority within the space agency. But in 2008 circumstances changed. President George W. Bush decided the space shuttle should soon retire, leaving a gaping hole in the agency's pipeline to supply the station. The backup plan moved to the front burner.

As a result, NASA accelerated the process of awarding contracts for actual supply missions. Although SpaceX, with Dragon, and Orbital Sciences, with its Cygnus vehicle, won COTS development contracts, NASA had no requirement to pick them for the operational phase of the program. The agency initiated an open competition called Commercial Resupply Services. For this program, NASA would pay one or two providers more than a billion dollars each—life-saving money—for a set number of supply missions.

That summer SpaceX could not be confident about winning. Despite the COTS award in 2006, Shotwell said most of her industry peers expected her company to fail at building a large orbital rocket. And after the second and third failures of the Falcon 1 rocket, those opinions did not change. A handful of NASA engineers like Mike Horkachuck, who had worked closely with SpaceX over the previous two years, were becoming believers. Yet plenty of doubters remained inside the agency and on Capitol Hill as the bid process heated up.

"NASA had worked with us for two years, and I think they were pretty pleased with us," Shotwell said. "But they also had some concerns. NASA was most worried at that time about our software, and the Flight Three failure was probably not helpful on that front."

But as summer turned into fall, SpaceX started to succeed. Flight Four reached orbit. In November, the propulsion team performed a full-duration test-firing of the Falcon 9 rocket in Texas. All of a sudden, it appeared as though SpaceX just might be able to build rockets.

Still, Musk fretted about how his company's desperate finances must look to NASA as his personal wealth dried up amid the recession. Musk also feared NASA might choose to award the contract to a single company instead of two. If that happened, SpaceX might well be left out. Orbital Sciences' new senior vice president for commercial space transportation, Frank Culbertson, had deep ties to NASA. After three space missions, the former astronaut served in management at NASA, and retained close relationships with agency officials making the selection. Orbital's headquarters were in Dulles, Virginia, and Culbertson was seen frequently that fall in the offices of key decision-makers at the agency in Washington, D.C.

Finally, on a Monday morning, December 22, 2008, the answer came. "They just called my cell out of the blue, right before Christmas," Musk said. NASA's human spaceflight chief, Bill Gerstenmaier, led the call. The leader of the International Space Station program, Michael Suffredini, was also on the line. They were excited to tell Musk that SpaceX had won one of two contracts. Musk couldn't believe it. He told them, "I love NASA. You guys rock." After the call, Musk asked Shotwell to immediately sign whatever deal NASA offered. He harbored a niggling fear the space agency might take the contract back. Two days later, on Christmas Eve, at 6 P.M., Tesla closed a financing round that provided the strapped automobile company with six months of funding. At a stroke, his two seemingly doomed companies were saved.

"It felt like I had been taken out to the firing squad, and been blindfolded," Musk said. "Then they fired the guns, which went click. No bullets came out. And then they let you free. Sure, it feels great. But you're pretty fucking nervous."

For Shotwell, the CRS award represented a huge win. Through the COTS contract, and now CRS, she had delivered two government awards that elevated SpaceX from a small start-up to a maturing company; from dozens to many hundreds of employees; and from the Falcon 1 to a powerful, world-class rocket. NASA had offered up the funding, and among dozens of suitors, Shotwell had brought it home. She saved SpaceX.

Not surprisingly, Musk offered Shotwell a promotion that fall. He had gone with a conventional hire for the company's first president two years earlier, choosing the seasoned aerospace leader Jim Maser. That experiment had failed. Perhaps, Musk reasoned, the best person for the job already worked for him. So he asked Shotwell if she wanted to manage more than just business development and legal affairs. And in December of that year, she became president of SpaceX.

"That was a good year," she said. "I look at 2008 very favorably. Elon looks at 2008 like just a horrible year in his life. But not me."

The Falcon 1 launch team returned to Kwaj in the summer of 2009 to fly the rocket's first dedicated commercial payload. The Malaysians had stuck with the company through six years and three failures. And now, the four-hundred-pound Earth-observing spacecraft would get its ride to space.

The launch on the afternoon of July 14 proceeded without incident. SpaceX hired Roger Carlson, a physicist from Northrop Grumman who had worked on the James Webb Space Telescope, to direct launch operations from Kwajalein. Following the flight, Tim Buzza stood with Carlson on one end of Omelek, talking about the future of the company. Four years had passed since Buzza and about two dozen engineers and technicians had come there to build a launch site from nothing.

"Rog, you own this island now," Buzza told the new site director.

"I'm going to Florida to do Falcon 9, and you just keep launching Falcons from here."

For a while, it seemed like that would happen. In early September 2009, SpaceX announced a contract to launch eighteen satellites for the U.S. telecom firm ORBCOMM. This campaign would entail multiple launches of an upgraded version of the rocket, called the Falcon 1e, with a larger first stage and an upgraded Merlin engine. The Falcon 1 rocket's first new contract in years, and one for multiple launches at that, augured well.

But a few weeks later, everything changed. Musk called a meeting of the Falcon 1 team, and told them, without preamble, that the booster had flown for the last time.

"That was tough for a lot of us that had worked the Falcon 1 missions," Chinnery said. "We spent so much effort and time on making that program successful, and to move on like that was classic Elon. He was very focused on what he wanted, and Falcon 1 wasn't in the plan beyond just learning how to do it."

Once they got beyond their initial shock, however, the Falcon 1 team accepted the wisdom of Musk's decision. It meant less work, as they would not have to spend time developing, testing, and building the Falcon 1e. With the extra hours, they could focus on Falcon 9 and Dragon, which now represented the future. Ultimately, it worked out, as the ORBCOMM satellites would eventually fly into space aboard the larger Falcon 9 rocket. As for Omelek Island, a few SpaceXers remained through the end of 2009 to clean up the site. Everything had to be moved off. The Army told SpaceX the concrete needed to be chipped into little pieces, no larger than a golf ball. Nature and the coconut crabs soon reclaimed the tiny island.

And what did the military think about the Falcon 1's abrupt end? Through its Falcon project to develop a rapid, reusable launch, D.A.R.P.A. had funded a couple of early SpaceX missions, and supported technology development with grants. The Falcon 1 rocket ended up being the only small-launch vehicle funded by the military program to actually reach orbit. After a decade, the Air Force still had not yet found a replacement. So did the company's abandonment of the Falcon 1 make the military wary of the vicissitudes of SpaceX and Musk?

"I didn't see it as a problem," said Steven Walker, who served as program manager for the Falcon project during SpaceX's developmental years, and later led D.A.R.P.A. "They went on to pivot into the Falcon 9 rocket, which has changed military space for the better. They are launching costly military satellites for one-fourth of the cost of what the United States government was paying before they came along. I'd say we got our money's worth from the Falcon 1 program."

While a skeleton crew wrapped up operations on Kwaj, the majority of SpaceX's launch team set up operations at Cape Canaveral Air Force Station in Florida. Less than four years after they abandoned the main Air Force launch facility on the West Coast, SpaceX obtained a lease for a historic site on the East Coast to launch the Falcon 9. The company rebuilt a nearly five-decade-old pad previously used to launch Titan rockets.

Following lengthy technical reviews of the new rocket by NASA, the Air Force, and other stakeholders, SpaceX proceeded with a series of static fire tests during the spring of 2010. Finally, it received a launch date for the rocket: June 4. As a result, less than two years after the first successful Falcon 1 rocket reached orbit, the company rolled its newest booster to the launchpad. It dwarfed the Falcon 1. That rocket had stood 68 feet tall and weighed about 60,000 pounds. The Falcon 9 reached a height of 157 feet and weighed a staggering 735,000 pounds fully fueled. If the Falcon 1 rocket was a toddler, the Falcon 9 was Shaquille O'Neal.

SpaceX moved the Falcon 9 out to its refurbished pad on June 2, and a day later, a classic Florida seabreeze-driven storm rolled in off the Atlantic, dousing the exposed rocket in torrential rains. Following the thunderstorms, a launch controller noticed a lopsided radio frequency signal coming from the second stage. This represented a showstopper, so that evening Buzza, Musk, and Altan drove out to the launchpad to troubleshoot with launch engineers there. The rocket, which had been raised vertically for launch, was lowered into a horizontal position just above the ground for inspection. This would have taken a full day with the Falcon 1 on Omelek, but SpaceX had applied that lesson learned when designing the launch cradle for the Falcon 9 booster.

Upon reaching the launchpad, Musk delegated Altan to climb a ladder to where the Haigh-Farr telemetry antennas were located on the exterior of the rocket's second stage. Altan had done this kind of work so often on Omelek, in the JLG lift, and here he was again rising to meet a Falcon rocket. After removing the covers, Altan confirmed the water intrusion problem. Following a brief huddle, they decided to try drying out the antennas with a blow dryer. Up on the ladder, Altan waved the dryer back and forth until he thought they were as dry as they were going to get. All the while, Musk and perhaps fifteen to twenty people watched him from below.

"That is as exposed as you can get," Altan said. "Elon didn't really interfere or say anything, he just let me do my thing, which was to open it up, blow it out and heat it up, and then seal it with silicone sealant to survive the next day and launch."

After the rocket had been buttoned up, and Altan climbed back down the ladder, Musk approached his director of avionics. "You think it is good enough to fly tomorrow?" he asked.

"It should do the trick," Altan replied.

Assessing this answer, Musk looked deep into Altan with one of his penetrating stares, seemingly to determine whether Altan had responded under pressure with what his boss wanted to hear, or if he was being sincere. Musk must have liked what he saw, because he eventually responded with a simple "OK."

By then it was pretty late. Any hopes of a good night's sleep before a long and crucial day were dashed. At about 3 A.M., Buzza drove Musk back to the hotel in his rental car. During the long drive down the length of the Cape, Musk peppered him with questions. But not about the launch attempt later that day. Similarly to how his mind had wandered during the first Falcon 1 launch in 2006, Musk looked ahead toward what came next. He asked Buzza about the Falcon Heavy, and recovering the Falcon 9 rocket's first stage. Typical Elon, Buzza thought as he dozed off that morning.

The next day's launch was very nearly perfect. The primary goal of the test flight had been to avoid damaging the launchpad. Secondarily, the company hoped to reach orbit. If the rocket got that far, it aimed for an orbit thirty-five degrees above the equator. It did all of that and more. With an accuracy remarkable for a brand-new rocket, the second stage inserted itself into orbit at an inclination of 34.494 degrees. Thus the untested rocket had screamed off a pad hundreds of miles away, reached a velocity many times the speed of sound, and missed its target orbit by a mere 0.006 degrees.

That night, SpaceX partied on Cocoa Beach Pier, which extends eight hundred feet into the Atlantic Ocean. For eight years, the company had scrabbled to make ends meet, struggling to put rockets into orbit, and nearly dying on several occasions.

For Elon Musk and SpaceX's rapidly growing workforce that night, those failures were behind them.

Overhead, their rocket soared among the stars.

Below, waves dashed against the pier.

And right in front of those employees, and the company they loved, lay a dazzlingly bright future.

. . .

Months before the first Falcon 9 rocket lifted off from Florida in the summer of 2010, Thomas Zurbuchen placed a few bets with friends on the mission's success. His friends were happy to bet against the upstart company, which lacked the heritage of the industry's main players, Lockheed Martin and Boeing. After four tries SpaceX might have gotten a dinky rocket to space, critics said, but they weren't ready to roll with the big boys.

Zurbuchen knew better. The Swiss-born scientist had helped build and run a highly regarded graduate program in space engineering at the University of Michigan, and in the spring of 2010 *Aviation Week* asked Zurbuchen to write about talent development. As part of this exercise, Zurbuchen made a list of his ten best students based on academics, leadership, and entrepreneurial performance during the previous decade, and researched where they had ended up. To his surprise, half of the students worked not for the industry's leading companies, but at SpaceX. The results blew him away.

"That was before SpaceX was successful," said Zurbuchen, who in 2016 became the chief of science exploration at NASA. "So I interviewed these former students and asked, 'Why did you go there?' They went there because they believed. Many of them took pay cuts. But they believed in the mission."

In his article for the aerospace publication, Zurbuchen wrote about how SpaceX had succeeded in the battle for talent with an inspiring goal. "I was a little bit nervous about betting on the immediate success of Falcon 9," he wrote. "But in the long run, talent wins over experience and an entrepreneurial culture over heritage." Too often in the modern aerospace world, he added, bureaucracy, rules, and a morbid fear of failure "poisoned" the workplace.

Published two months after the inaugural Falcon 9 flight, the article caught Musk's attention. He shared it with all his employees, saying they were the best and brightest in the industry, and people were beginning to notice. Musk also invited Zurbuchen to SpaceX's factory for a tour. During the visit, Musk thanked Zurbuchen, and they discussed the company's legion of doubters. But then, Zurbuchen recalled, Musk suddenly focused on the scientist with his characteristic, arresting stare. The pleasantries, such as they were, had ended. Musk asked a single question: Who were the other five students?

"I realized that was what the whole meeting was about," Zurbuchen said. "The meeting was not about me. He wanted to recruit them. He wanted the other five."

Not everyone appreciated the *Aviation Week* article. Zurbuchen received phone calls after the piece that subtly, and not so subtly, intimated that he must be an egghead academic who had drunk the SpaceX Kool-Aid. He remembers angry conversations in hallways, and being told at meetings about his ignorance of the launch business. But Zurbuchen stood by his conclusion. When he spoke with engineering peers at places like the Massachusetts Institute of Technology or the University of Southern California, he heard similar things. SpaceX had juice with their students, too: the freedom to innovate and resources to go fast summoned the best engineers in the land.

Competitors began to take note of SpaceX's success, too. The Falcon 1 had been a minor annoyance to most U.S. aerospace companies, really threatening only Orbital Sciences and its Pegasus rocket. But the Falcon 9 represented a legitimate challenge to the industry's power brokers.

United Launch Alliance enjoyed a monopoly on U.S. national security launch contracts, and a handful of large aerospace firms, including Boeing, Lockheed Martin, Aerojet Rocketdyne, Northrop Grumman, and ATK Aerospace carved up most of the rest of the launch business for other government contracts, including NASA. None of these companies welcomed a new competitor, especially one so potentially disruptive. In response, they started to fan the flames of political opposition. Just as these contractors had a vested interest in the status quo, so did politicians in Alabama, Florida, Texas, Utah, and a handful of other states with a disproportionate number of aerospace jobs.

SpaceX took off at a key moment in space policy history. In 2010 a battle raged between the White House and Congress over the future of human spaceflight. With its final mission set for mid-2011, everyone agreed the space shuttle would soon be retired. The big aerospace companies, which all held large contracts for the shuttle program, helped Congress devise a plan to continue on similarly lucrative contracts to build new government spacecraft and rockets. The Obama White House sought to limit funding for these expensive programs, and give emerging players like SpaceX a chance to see if they could bring down spaceflight costs. The first Falcon 9 launch, then, offered something of a referendum on President Barack Obama's space policy. If the rocket failed, the naysayers would be justified in their view that commercial space was not ready for prime time.

"I was well aware that not only my own reputation, but the success or failure of the Obama administration's space policy, would be largely determined by the outcome of the SpaceX launch," said Lori Garver, NASA's deputy administrator at the time and a key Obama space advisor.

One might think members of Congress would welcome the addition of a new, American-made rocket to the nation's fleet. At the time, most of the country's military assets launched on an Atlas V rocket, which was powered by Russian-made engines. But political space leaders on Capitol Hill, allied with the existing aerospace power brokers, offered a muted response. The senior senator from Texas, Kay Bailey Hutchison, said, "Make no mistake, even this modest success is more than a year behind schedule, and the project deadlines of other private space companies continue to slip as well." Such a tepid reaction is remarkable given that

SpaceX had established the large, and growing, McGregor test facility in Hutchison's home state.

Much of the space agency's leadership likewise watched the Falcon 9 rocket's ascent with a sense of wariness. Outside of Garver and a few allies, who saw lowering the cost of launch as the next crucial step for the agency, resentment lingered over the Obama administration's efforts to privatize part of the space program. Some of these decision-makers honestly believed SpaceX was too brash and its approach too risky. But the reality is that a lot of NASA leaders also had ties to the legacy aerospace companies. Then, as now, senior officials at NASA often came and went between the agency and large contractors. This revolving-door effect helped the aerospace industry maintain some control over the direction taken by the agency, and reinforced skepticism of companies like SpaceX that sought to shake up the existing order.

But shake things up, SpaceX would. After the Falcon 9 rocket's initial success, SpaceX would prove its Cargo Dragon spacecraft could fly safely. Six months after the first Falcon 9 launch, the rocket's second mission lofted Dragon into space for the first time. As an homage to Monty Python's iconic sketch "The Cheese Shop," the capsule carried a wheel of Brouère cheese. It is amusing to think that the company's first orbited payload was RatSat, and its first Dragon spacecraft carried a cheesy comestible. Three hours after its launch, Dragon splashed down safely in the Pacific Ocean. Whimsical payload aside, no private company had ever flown and landed a spacecraft before. Then, in May 2012, Dragon docked to the International Space Station for the first time. It has since flown twenty successful cargo missions into space.

Booster reuse was part of Musk's plan from the beginning. For all Falcon 1 launches, SpaceX not only stuck a parachute at the top of the rocket's first stage, it dispatched a ship with an employee to retrieve flight hardware from the ocean. In 2006, the Falcon "Recovery Team" consisted of structures engineer Jeff Richichi and an eighty-foot Army boat named *Great Bridge*.

Stationed about ten miles outside of the area where SpaceX calculated the first stage would land in the water, the slow-moving *Great Bridge* began plodding toward the splashdown zone immediately after the Flight One launch was confirmed. Sea-based communications being what they were, the boat's crew did not learn right away there was no rocket to recover.

Prior to the launch, SpaceX had published this potential area of impact in the Merchant Shipping Notices. When the Army vessel arrived at the splashdown point, Richichi was surprised to find a Chinese boat waiting there, purportedly fishing. This seemed like much more than a coincidence.

"In the hundreds of square miles of open ocean in which to go fishing, this trawler was in the exact location of the splashdown point, within two hours of the splashdown time," he said. "I'm sure this had nothing to do with the fact that there was a first-stage rocket expected to come down there, via a parachute, at that time."

Later, SpaceX would come to realize that it had no hope of capturing a first stage after it fell back through Earth's atmosphere, and deployed a parachute. Reentering at supersonic speed, it would burn up long before the chute could deploy. But at the time, Richichi felt pressured to reel in a first stage. This was an all-but-impossible task as, after launch, he had to scan the skies and seas for a white, city-bus-sized object falling into an ocean full of white caps more than fifteen miles away. During the second mission he made the mistake of telling the crew of the *Great Bridge*, as an incentive, that he would give $100 to the first person to spot the first stage. The problem was that someone called out a false sighting of the first stage every minute or two.

"It was a nonstop callout of false sightings," Richichi said. "We were zigzagging all over the place, chasing first-stage ghosts. It was a terrible idea on my part, and I didn't try that stupid trick again."

It says something about Musk's commitment to reuse that he traded invaluable mass on the Falcon 1 by installing a parachute in the forlorn hope of recovering a first stage. His argument for reuse is simple: If an airline discarded a 747 jet after every transcontinental flight, passengers would have to pay $1 million for a ticket. Similarly, if every rocket flown into space drops into the ocean, space will remain cost prohibitive for all but a few wealthy nations and a few exclusive astronauts. To make humanity a multiplanetary species, Musk sought to lower the cost of getting into space and flying onward to other worlds.

The early returns on the reuse experiment were, nonetheless, sobering. "We were very naive at the time, but the expectation was that we would throw a parachute on this thing and recover it," Musk said. "We were huge idiots."

SpaceX never got there with the Falcon 1 rocket. And it took a lot of mucking about with the Falcon 9 rocket, too. During its first launch in 2010, the first stage broke up during reentry. The company later recovered pieces of the booster, including helium pressure tanks, the drogue parachute, and the housing for one of the engines. When Musk says they were "huge idiots," it's because the company's engineers didn't really grasp the futility of putting a parachute up against the kind of reentry energy carried by a multiton rocket moving at several times the speed of sound.

To make it all work, SpaceX needed to provide some sort of a heat shield to protect the rocket as it screamed back down through the atmosphere. Still more critically, they needed to master a technology NASA had studied only in simulations and wind tunnels. To control the Falcon 9 rocket and slow it down, SpaceX had to relight the rocket's engines high in the atmosphere when the booster is traveling at Mach 10. Understandably, a lot of engineers were concerned about the stability of the first stage during this turbulent period, when its rocket engine would ignite and fire directly into an atmosphere rushing toward it. SpaceX began testing this technology, called supersonic retropropulsion, as far back as September 2013. Finally, the reentry team had to come up with a mechanism to steer the booster through the thickening atmosphere, toward a landing site. If the goal was to rapidly turn a booster around and fly it again, dropping it into the ocean was probably not a good idea. The company had learned its lesson about saltwater corrosion with Flight One.

It took a good bit of tinkering and failure, but on the twenty-first flight of the Falcon 9 rocket in 2015, the company brought its booster to a safe nighttime landing at a brand-new pad at Cape Canaveral Air Force Station in Florida, just a couple of miles from the launch site. After the nighttime launch and landing, just three days before Christmas, employees at the company's factory in Hawthorne cheered raucously and then broke into chants of "U-S-A, U-S-A!"

Musk was thrilled. "I wasn't at all confident that we would succeed, but I'm really glad of it," he said that night. "It's been thirteen years since SpaceX was started. We've had a lot of close calls. I think people here are overjoyed."

Prior to landing along the Florida coast, SpaceX had experimented extensively with putting boosters down on an automated drone ship stationed offshore in the Atlantic Ocean, downrange from the launch site. This is simple physics. Shortly after a rocket takes off, it leans forward, gradually moving from a vertical position with respect to the ground into a horizontal posture, setting up for an orbital insertion. The net effect of this is that by the time it separates from the second stage, the first stage is happily zooming away from the launch site at a very high speed. This is not good if you want the booster back. To return the rocket all the way to a landing site in Florida requires a long burn of the first-stage engines, and this takes a lot of fuel. Any fuel used during the return trip cannot be used ascending, so this represents a huge penalty to the mass a rocket can send into orbit. One solution to this is sending a boat hundreds of miles offshore, to catch the rocket downrange.

Only it's rather hard to land a rocket on a boat that is bobbing up and down in the ocean. It requires some damn good computer programming to make the rocket and autonomous drone ship line up just right, and no one had ever done it before. Until it happened. On April 8, 2016, a Falcon 9 rocket launched a Thai communications satellite toward a high orbit around the Earth, and then, as if it were a magical being, the first stage landed on a drone ship playfully named *Of Course I Still Love You.* It was one of the most jaw-dropping things I have ever seen. For the first time in my life, I felt as though I had seen something as cool as my parents had witnessed during the Apollo Moon landing in 1969.

And then they did it again. And again. Suddenly, SpaceX had a hangar in Florida full of first stages. "It even surprised us that suddenly we had ten first stages or something like that," Koenigsmann said. "And we were like, well, we didn't really account for that."

Today it is normal for SpaceX to launch rockets, catch them on land and at sea, and fly them again a couple of months later. In fewer than three years, the paradigm has shifted entirely. Whereas it once seemed novel to reuse a rocket, now it seems almost wasteful to throw them away. The company's competitors initially scoffed at the notion of launching a rocket vertically, landing it vertically, and then flying again within a few months. Now they're scrambling to catch up. State rocket institutions in China, Russia, Japan, and Europe have all funded reusable development programs at some level. It's the same for rocket companies seeking to compete with SpaceX in the United States, including Blue Origin and United Launch Alliance.

SpaceX did not stop there. In 2018, the company launched the Falcon Heavy for the first time, giving it the world's most powerful rocket. Essentially, this titanic booster strapped together three Falcon 9 first stages to form a monster first stage. In less than a decade, the company went from launching a rocket with a single engine to one with twenty-seven engines. The world had never seen anything like this before. And then, the two side boosters returned to Earth, landing side by side, almost as a pair of synchronized swimming angels descending down from heaven to Earth.

Even President Donald Trump, so often consumed by his personal interests and the politics of the moment, took note of the Falcon Heavy's graceful launch and landing. "You see the engines coming back down, there's no wings or nothing," he said at one campaign event. "It's almost like, what are we watching? Is this fiction?"

It sure seemed like science fiction, though it was not. Nor was the company's rewriting of the global launch industry. In the mid-2010s, SpaceX began to deliver on the promise of low-cost, rapid launch. At about $60 million for a basic Falcon 9 launch, the company undercut every other major orbital rocket on the market. Commercial satellite operators who had long looked to Europe, Russia, or China to get their large birds into space suddenly started flocking back to the United States for the first time in decades. SpaceX grabbed about two-thirds of the world's commercial satellite launch business by the end of the decade. Some large fleet operators spread their business around just to make sure SpaceX's competitors would not go out of business.

With the Falcon 9 rocket, SpaceX succeeded where the Falcon 1 ultimately failed in its quest to find a diverse array of customers. The Falcon 9 is powerful enough to capture a large share of the commercial satellite market, as well as science missions for NASA and military payloads for the Air Force. SpaceX has also won cargo and crew contracts from NASA, and now has its eyes on deep space. From these profits SpaceX has been able to invest in Musk's ambitious Starship program, which he believes is critical to sending enough people and cargo to Mars to establish a self-sustaining settlement.

The company's success has shaken the aerospace world to its core. In 2016, the vice president of engineering for United Launch Alliance, Brett Tobey, gave a candid talk at a University of Colorado Boulder seminar. He did not know he was being recorded, but his remarks were later made public. In his talk, Tobey admitted that United Launch Alliance and its fleet of Atlas and Delta rockets, which until SpaceX came along had a monopoly on launches for the U.S. Air Force, had no hope of competing on price.

"We're going to have to figure out how to bid these things at a much lower cost," he said of contracts for national security launches. Tobey also admitted that his company's launches cost the government about three times more than SpaceX's prices. He was gone from United Launch Alliance within days, but in reality Tobey was just saying what everyone in the industry already knew. In just its second decade, SpaceX had disrupted the global launch industry.

Finally, and perhaps most significantly, SpaceX legitimized the new space ethos of lowering the cost of access to space. It has shown that private companies and private capital, working alongside the government, can do amazing things in space. Entrepreneurs have had an easier time attracting funding for all manner of space ventures after investors witnessed the success of SpaceX with its Falcon 1 and Falcon 9 rockets.

"It helped the whole industry," said Peter Beck, whose successful Rocket Lab company has launched more than a dozen small Electron rockets from New Zealand since 2017. "It proved that a private company could successfully deliver cargo and satellites to orbit. And not just for launch, but for spacecraft as well, they showed that a commercial company can play in a domain that had typically belonged to the government."

Nearly two decades have now gone by since Musk first began thinking seriously about Mars. During an interview in early 2020, his mind drifted back to that first impulse to get into the space business. He remembered a gray, rainy day on the Long Island Expressway with his friend Adeo Ressi, and later his frustration upon visiting NASA's website and finding no plans. He could not understand why humans had remained stuck in low-Earth orbit since Apollo. And so he made a life-changing decision to commit himself to the goal of Mars, a commitment that has grown stronger over time.

"That's nineteen years ago, and we're still not on Mars," he said.

"Not even close," I replied.

"Yeah," he agreed. "Not even close. It's a goddamn outrage."

This is the passion that fires Elon Musk, and impels him to drive his teams forward every single day. Decisions in his world come down to a simple calculus: Will this get humans to Mars sooner, or not? Little else matters in his mind. Though we are not yet close to Mars, we are leaps and bounds closer today than ever before. Musk's first step was to bring down the cost of launch. Against all odds, he has done so. Now his company, with Musk's constant urging and the accumulated knowledge of the last twenty years, is building the Starship vehicle to one day carry settlers to Mars.

SpaceX has undeniably come a long way from its scrappy beginnings in El Segundo, and the desperation of trying to launch from the nearby hills north of Los Angeles. Those were madcap days, first running out of LOX, and then running into red tape, and finally running to Kwaj. But what started there changed the world. One day, perhaps SpaceX will change another world, too—by transforming Mars from a lifeless red planet into a living green Eden.

EPILOGUE

I was fortunate to speak with dozens of current and former SpaceX employees, often at great length, for this book. Through their recollections I have tried to bring the story of SpaceX to life, and provide a sense of the personal sacrifices that went into the making of this great American rocket company. I am indebted to all of these people for their time. By way of closing this book, I find it appropriate to offer a few final thoughts from some of the most important participants, along with their titles during the Falcon 1 heyday, and an update on their activities since Flight Four.

CHRIS THOMPSON, Vice President of Structures

Musk placed enormous demands on his employees. About half of those who survived the Kwaj crucible remain with SpaceX today, but the other half have moved on, often to escape the burden of toiling under Musk after it became too much. Many transitioned on to other start-up rocket companies, chasing the thrill of working with a small band of fellow-minded warriors, and striving to build something that could, however improbably, blast off the surface of the Earth and break the bonds of its gravity.

Chris Thompson said he effectively quit in early 2008. He had made the ninety-minute drive from his home in Costa Mesa for nearly six years, and it all just became too much. "I had gotten to the point where I was tired of commuting, I missed my kids, I was spending way too much time at work, my wife was on the ragged edge, and I basically just felt I had to make a change," said Thompson. He left to work at a company founded by John Garvey, his friend and one of the rocket scientists who had advised Musk before he launched SpaceX. As Thompson remained the driving force behind the company's structures department, and with Falcon 9 development ramping up, a couple of the other vice presidents stepped in to encourage Musk to keep Thompson on part-time. Musk agreed.

Only five months into the venture with Garvey, it began to run out of money. So Thompson swallowed his pride, and sent an email to Musk, asking for his full-time job back. And then he waited. For a day. A week. And then three weeks. There was no response from Musk, who typically made brisk replies by email. Thompson figured he had burned his bridges at SpaceX. But then, Musk called him.

"Hey, I saw your note," Musk said, as if three weeks had not slipped by in the interim. "Why don't you just come in on Monday? You can have your old job back. See Jerry Fielder and he'll give you a whole new salary."

Thompson recalls being stunned. He stammered something along the lines of "What???" before Musk hung up. The whole call lasted only a minute or two. When Thompson returned to SpaceX the next Monday and met with Fielder, the head of human resources, Thompson had received a raise and more options. Musk treated Thompson like nothing had ever happened.

He remained at the company for four more years, ordering the Falcon 1 rocket stripped like a Chevy, enjoying the success of it launching, and seeing the Falcon 9 through its first couple of missions. But by then the excitement had faded as the company moved into operations as much as development. During the first half of the 2010s, the primary goal at SpaceX entailed reaching a point where the Falcon 9 rocket could be launched frequently, at least on a monthly basis, to fly out a backlog of missions for customers who had flocked to the low-cost rocket. Thompson's relationship with Musk deteriorated, too. As Thompson stood up to Musk, he felt the boss's wrath more and more. They had shouting matches. It got to the point where Thompson didn't need any of that any longer.

It had all been quite a ride and made Thompson a wealthy man, but the SpaceX experience came with a high personal cost. Thompson had just turned forty years old when he started at SpaceX. His kids were coming into some of the most important childhood years. A son, Ryan, was twelve at the time. A daughter, Taylor, was nine. Over the course of the following decade at SpaceX, Thompson would miss the bulk of their teenage years. All the while his wife Susan had worked a full-time job, too.

"That was hard stuff," he said. "There was no work-life balance. The impact was, you don't see your kids. You miss out on parent-teacher conferences, you miss out on plays, you miss out on soccer matches, you miss out on baseball, you miss out on volleyball, you miss out on a lot of those things that are important in a kid's life during their formative years."

And so Thompson began to prioritize family over his profession. He departed SpaceX in May 2012, and, after a brief stint at Blue Origin, settled in at Virgin Galactic, which at the time had begun developing a small rocket that could be dropped from a modified 747 and launched from high altitude. It helped that chief executive George Whitesides made it clear that eighteen-hour days were not the norm. Thompson remained there for five years, before leaving to head up engineering for Astra, a secretive launch company designing a no-frills, small satellite launcher. Thompson said he loves it there. But that is a story for another day.

BULENT ALTAN, Responsible Engineer for Avionics

After Musk reached out to his friend Larry Page, Altan and his wife had moved to Los Angeles in 2004. Over the following decade, Altan had enjoyed the adventure of a lifetime. Before the first flights of both the Falcon 1 and the Falcon 9, Altan had found the eyes of the company on him as he worked in the sky, performing surgery to first a capacitor on Kwajalein, and later water-damaged antennas in Florida.

He had joined SpaceX because of the daring of Musk and the boldness of what SpaceX aspired to do. "I didn't go to school just to go to meetings, and sit in a cubicle to try and perfect a single screw," he said. "This was a company that wanted people to just get stuff done. I wanted to get my hands dirty, and no other company was really going to offer that besides SpaceX."

He got his hands dirty. On his very first day on the job, Altan designed a printed circuit board and sent it to manufacturing. He mused, at the time, that at most other companies he probably would not even have had an IT account set up by the end of the first day. Soon, he was building rockets, fixing them, and even cooking for his fellows at night on Omelek Island. His Turkish goulash proved so popular that he wrote down the recipe, as if it were a launch procedure, and shared it with his colleagues. Altan would go on to lead the avionics department at SpaceX, before leaving the company in January 2014. On his last day, the cafeteria in Hawthorne served his goulash to employees.

He returned to SpaceX in 2016 for two more years. With his

programming skills, Altan served as a senior engineer on the company's new Starlink project. This is SpaceX's ambitious plan to put thousands of small satellites into low-Earth orbit, and provide global internet service. To make this work, the satellites must communicate with one another as they zip overhead, creating a seamless stream of data for users on the ground. Altan left SpaceX just before the first prototypes were shipped. He cofounded a venture capital fund.

If, one day, you get your internet from space, among the people you can thank is the guy who overcame his fear of heights—and also happens to make a mean Turkish goulash.

ANNE CHINNERY, Operations Manager

She had more launch sites to help set up after Kwajalein. Five years after SpaceX made a hasty exit from Vandenberg, the Air Force agreed to allow the company back on the property. Chinnery helped design and develop the site where a Falcon 9 rocket would launch from the West Coast for the first time in 2013. She also worked on setting up a vertical launch facility at the company's test site in McGregor. There, the company performed a series of landing tests to demonstrate the rocket's ability to hover and move laterally above the ground that would ultimately lead to first-stage landings and reuse.

But by the end of 2013, after more than a decade at SpaceX, Chinnery had lit every match in her book. She had nothing else to give. During those early years, the work had stimulated her. It had been hard, exciting, and rewarding. But quietly, it took a toll. "Because it was never boring, it was easy to lose track of the fact that you were getting tired and stressed," she said. "It was just so fun that you always wanted to keep going back and doing more of it."

Even the trips to Kwaj became burdensome. The surroundings were beautiful, but Chinnery and the others worked so hard, they could only rarely enjoy the beaches, crystal clear water, and sunshine.

"That chronic work level and stress, eventually, does a number on everybody," she said. "I definitely had my moments out there, and the chronic stress of working for SpaceX over the course of eleven years pretty much turned me into a basket case by the time I was done. I was able to recover from that, but only after a couple of years away from SpaceX." Chinnery has no regrets. She cherishes the time spent at SpaceX: "It was the experience of my life."

By the summer of 2015, she was ready for a new chapter. Chinnery started working for Tom Markusic at Firefly, a small rocket company. She marvels at the difference between the military's attitude between then and now. To entice Firefly to launch from Vandenberg, Air Force officials have eagerly helped, and proactively found solutions to problems.

"There is just no question that without SpaceX it would not have happened," she said. "They convinced everybody else that commercial space is a real thing. And when they did that, the Department of Defense realized that they could either be a part of that, or they could be left behind."

So many people like Chinnery felt burned out by their years at SpaceX because Musk pushed them relentlessly. His schedules were invariably aggressive. Time was money. This window to reach Mars and make humanity a multiplanetary species, Musk fears, may not remain open forever. And Musk's own lifetime was finite. This brutal devotion to speed got results. The first Falcon 1 launch attempt came a mere three years and ten months after Musk started SpaceX. The company reached "space" in four years and ten months. It made orbit in six years and four months. It did all of this starting with just three employees, limited government funding, and building a rocket from scratch with mostly in-house components for the engine and rocket.

These timelines are all the more impressive compared to the second wave of small satellite launch companies like Firefly. Dozens have been created since SpaceX started firing rockets into space. One might think these companies would have it easier. Launch sites want them. SpaceX demonstrated that private capital can do meaningful things in space. And regulators have learned from SpaceX what commercial launch is all about, and they have a political mandate to help rather than hinder. But the new companies have gone slower.

Only one private company with new technology, Rocket Lab, has actually reached orbit. It took eleven years and seven months to do so. Firefly was founded in January 2014, and as of Fall 2020 had not reached orbit or even attempted a launch. Virgin Orbit began to get serious about building a small orbital rocket in December 2012, and it, too, had not reached orbit by late 2020. Blue Origin, SpaceX's most prominent new space competitor, was actually founded earlier, in 2000. It has taken a more stepwise approach, but for all of Jeff Bezos's money has yet to launch a rocket into orbit after twenty years.

Chinnery said she believes companies today are more cautious due to the changing market. When Shotwell began selling the Falcon 1 rocket, customers were desperate for cheaper small-launch service. Today, there are half a dozen well-capitalized companies with strong technical plans. Customers can afford to wait and see who succeeds before signing on, and they're much less tolerant of risk.

"That's one of the things Elon brought to SpaceX—risk tolerance," Chinnery said. "He didn't want to fail, but he wasn't afraid of it. And I think in a lot of other aerospace businesses, there still is that fear of failure, they want to be better than that."

With so many competitors, companies today really can't afford to fail. So they go a little bit further. They test hardware more. If one of many models were to show significant sloshing in the second stage of their rocket, they wouldn't chance it. They would take the extra time to understand the problem. For if Chinnery and Firefly fail once or twice with their Alpha rocket, there probably won't be a third or fourth chance.

TIM BUZZA, Launch Director

He hung around to see the Falcon 9 rocket fly several more times after that first flight in Florida, including chaperoning its first major upgrade, and launching it from Vandenberg.

By then, however, he felt as though the next generation had come through. Some of those young engineers who joined the company right out of graduate school, like Zach Dunn, Ricky Lim, Flo Li, and Tina Hsu, had grown into senior leadership positions. They had acquired the DNA from the earliest days of SpaceX, in an empty building in El Segundo, and were now rapidly spreading it to the next generation. To the scrappiness of early SpaceX they had added the success of a maturing company.

"For the original DNA, I think Elon has to be the key," Buzza said. "I don't think any of this happens the way it does without Elon. That's 100 percent. But then also I think that having people like Tom Mueller and Hans Koenigsmann and Chris Thompson and myself was important. We brought some heritage aerospace experience, but also were willing to be totally molded by Elon to change our thinking."

This had not always worked, such as with Jim Maser, who also had heritage experience. But the vice presidents who joined SpaceX from industry early on recognized that Musk's in-your-face management style came with benefits. He empowered his people to do things that would have required committees and paperwork and reviews at other companies. At SpaceX, if they could convince the company's chief engineer of something, they also earned approval from the chief financial officer, as they were one and the same.

The two big NASA contracts—first the COTS agreement in 2006, and then CRS at the end of 2008—catapulted the company to new heights. Early on, when SpaceX had perhaps 150 people, Musk put heavy restrictions on head counts and resources. This meant those early employees had to do the jobs of three or four people, and work brutal schedules. It meant that Buzza would read his children bedtime stories over the phone one night from Texas, fly home for a few days, and then spend the next two months in Kwajalein for a launch campaign. The NASA funding changed all of this.

"I don't want to diminish any of the accomplishments since Falcon 1," Buzza said. "They're unbelievable. But they certainly got the money and the resources to allow it to not just stay linear, but to accelerate things."

And as ever, Musk remains the dominant driver behind the acceleration. While trying to get Falcon 1 flying, he wanted specs for the Falcon 5. Then his small company took on the challenge of building both the Falcon 9 rocket and Dragon simultaneously. In the mid-2010s, as the company was well on its way to developing arguably the world's best rocket in terms of price and performance, Musk pushed for rapid reuse, and then the Falcon Heavy, and a Starlink internet constellation as well as the Starship and Super Heavy Launch System.

This incredible pressure wore down his employees, but for someone like Musk who sees only a narrow window to execute his sweeping vision, there is no other way.

"Sometimes it would kind of get under your skin," said Buzza, recalling Musk's focus on the Falcon 5 rocket in the Kwaj control center, during the final countdown to that very first launch. "Like, I'm over here struggling to solve the Falcon 1, and you're bugging us about a Falcon 5. But if you don't have somebody like that pushing the train out to the future, your iteration time is too slow. It's just too slow."

Buzza left SpaceX in mid-2014 to join Thompson at Virgin. After four years he moved to the rocket company Relativity Space, where he serves as the company's distinguished engineer. Relativity is the spiritual successor to SpaceX in terms of boldness. It seeks to 3D-print the entirety of its rockets to speed development and bring down costs. One day, it aspires to 3D-print a rocket on Mars, and launch from the red planet. Buzza, no doubt, will play an essential role in that, as he has done with this book. Many of the stories in these pages began with a tip from Buzza. Many of the details came from his notes and timelines. He has answered so many of my questions and made this story so much more authentic. If you've enjoyed *Liftoff*, you owe Tim Buzza a beer. I owe him many.

TOM MUELLER, Vice President of Propulsion

The relentless pressure finally got to Tom Mueller in late 2013. For a dozen years, he had worked long hours and weekends to make Falcon rockets go. Meanwhile, his daughter was growing up, and he was missing out. Stress from those years would eventually contribute to a divorce. "It was a critical time, and I just wasn't around much for them," Mueller said of his family.

The job had changed, too. Over the years at SpaceX Mueller brought three versions of the Merlin engine to flight, and felt he had very nearly reached perfection with the Merlin 1D, which powered the Falcon 9 rocket from 2013 onward. Through lessons learned and better technology, such as a more efficient turbopump, the final Merlin produced more than twice as much thrust, 192,000 pounds, as the Merlin 1A, with 76,000 pounds. But by 2013, much of this work was done. Production differed from development, and Mueller wearied of late-night calls about supplier issues as the Hawthorne factory cranked out more and more rockets with nine engines each.

"I'm like, you know, this is not what I'm good at," Mueller remembers thinking at the time. "I develop engines. So I told Elon I wanted to start to step down. Gwynne was there at the meeting, and Gwynne was horrified."

Through her sales experience, Shotwell understood the importance of Mueller to the SpaceX brand and, more important, its home-grown rocket engines. She and Musk persuaded Mueller to stay on for three more flights of the Falcon 9 rocket with the new Merlin 1D engine, figuring this would convince satellite operators that the supposedly new-and-improved propulsion system had actually improved. Six months later, Mueller went back to Musk and asked again. Musk realized his chief of propulsion was serious. So they came up with a plan where Mueller would get a new title, chief technology officer.

"It was bullshit," Mueller said. "But it was a great title, so it didn't look like I was taking a step down." As his workload eased, Mueller's physical health improved. He had planned to have surgery on a pinched nerve in his neck, due to the stress. But after he stepped down, he canceled the surgery.

The onetime logger from Idaho had always liked fast things, and he raced cars as a hobby. As SpaceX's propulsion chief, whenever Mueller got out of his Porsche after a race, he had to check his phone. Often, Musk had called with some problem or other demanding a response. "My pit boss was always like, 'Mueller, if your head's not in the game of racing, you shouldn't be racing,'" he said. "I'm like, 'No, no, I'm here.'"

Despite the sacrifices made in his personal life, however, Mueller harbors few regrets about his time at SpaceX. Since Mueller departed the company, the Merlins have been enhanced by Musk and the propulsion team he left behind. But the fundamental design remains the same. Once a month, or more, he has the pleasure of seeing the engines he designed launch rockets into space, and then carefully guide them back to Earth. The Merlin 1D engine powers both the superefficient Falcon 9 rocket as well as the world's most powerful booster, the Falcon Heavy. In May 2020, nine Merlin engines launched NASA astronauts into space from the United States for the first time since the space shuttle retired, ending a hiatus of nearly a decade. Mueller watched that mission, tense. He'd done the original design for the Dragon spacecraft's sixteen Draco thrusters and eight SuperDraco thrusters, the Merlin vacuum engine in the upper stage, and the nine Merlin 1Ds in the first stage—thirty-four engines in all. Now, for the first time, human lives were on the line. The engines burned hot and true.

Four days later another Falcon 9 rocket launched a bunch of satellites. This was the *fifth* time this particular first stage had gone into orbit. These days, his Merlin engine just flies, and flies, and flies some more.

"I'm extremely, extremely proud of what we've done," he said. "The Merlin 1D is such an awesome engine. I'm super proud of it. I can't take much credit for Raptor. I designed the original Raptor, but it has changed so much that I can't take much credit. I named her, and I hired and trained the team that developed Raptor. That's the credit I'll take. But the Merlin 1D was my baby."

ZACH DUNN, Responsible Engineer for Propulsion

Dunn got to SpaceX as soon as he could, and gave everything he had. He arrived just as Jeremy Hollman was looking for an exit, and he soon found himself working side by side with Mueller, already a legend in the industry. After his Flight Four heroics, Dunn continued to grow with the company, assuming a multitude of leadership roles in propulsion and launch. When interviewed for this book, Dunn was senior vice president of production and launch for SpaceX.

Engineers who come to SpaceX generally understand they are going to be used up in the job. The work can be all consuming. But what critics of these work schedules fail to grasp is that most new SpaceX hires willingly sign on to this bargain. They *want* that golden ticket for the world's greatest thrill ride.

Dunn climbed inside an imploding rocket, within a C-17 jet, at twenty-five thousand feet, with the fate of his entire company in his hands. But his journey did not end there. A decade later, he was still dancing on the bleeding edge of the possible, be it through landing rockets on boats or building interplanetary Starships. SpaceX wants to go to Mars, after all, something no company, space agency, or country has ever done. Will SpaceX get there? Maybe not. But for the adventurous it sure beats pushing a pencil at a government job where things move slowly, and working large exploration programs subject to cancellation every time a new president moves into the White House.

Dunn would do it all again, in a heartbeat.

"Certainly, I put a ton in," he said. "And I put in these, like, awesome super-productive years of my life. But it was where I wanted to spend it. I gave everything I had, to the exclusion of girlfriends or anything. I freaking gave it everything, and I wanted to give it everything. I don't think it was a toll, it was a trade."

He made the trade until May 2020, when Dunn left to join Buzza at Relativity Space. He wanted to be part of a small, against-the-odds team again, and build hardware from scratch. And maybe, after nearly fifteen years of giving it all, he had less to sacrifice to the rowdy and righteous mission of SpaceX. Dunn also allowed that he looked forward to spending more time with his two four-year-old twins, Zohra and Theodore.

As much as anyone in this book, Dunn embodied the passion and heavy metal spirit of SpaceX. He rocked out in McGregor atop the test stands, and played the front man on Omelek. And yet, although no one rocked harder at SpaceX than Zach Dunn, the band played on. Only a couple of weeks after Dunn left the company, the astronauts flying on board the first Crew Dragon mission blasted AC/DC's "Back in Black" on their ride to the launchpad.

FLORENCE LI, Responsible Engineer for Structures

She never thought much about leaving SpaceX. After working on the Falcon 1 all through the Kwaj years, Li moved on to the Falcon 9 launch program. She enjoys the sense of urgency at SpaceX because it feels like she helps to actually get things done. As SpaceX changed the world, Li felt privileged to play a part. She still has not given up her dream of traveling into space one day, but now feels like she serves a deeper purpose in working at SpaceX as an engineer to design and build things that go into space.

Like Buzza, Li partly attributed the company's success to Musk's hiring of an exceptional team of vice presidents who had both excellent technical skills and good leadership qualities. These leaders, and the young engineers like Li who worked for them, gelled during the Falcon 1 pressure cooker, which fed into the development of Falcon 9, and so on. But it was more than just hiring. Musk was there every step of the way.

"Having Elon, it makes things a lot simpler because he is super involved, he makes those difficult decisions," she said. "When those times came, he would step in and make those decisions like, *Are we going or not? What are we doing here?* And he's always kept us focused on that vision. He never would relieve us of any little detailed duties, but he would always make sure that we would come out and look at the bigger picture. I think that was really important to kind of keep that focus."

And then, the JLG Queen allowed, maybe SpaceX had a little luck along the way, too.

BRIAN BJELDE, Mission Manager

After Flight Four Bjelde stepped away from engineering. It was not that he was a bad engineer. Far from it. But Bjelde had tremendous people skills that proved useful in working with customers, writing proposals, and selling the SpaceX brand. Over time, Bjelde found that trading sunburns and chafed legs for sales and chai lattes suited him.

Musk liked what he saw, too. Even though Bjelde lacked a background in personnel and hiring, Musk asked him to become vice president of human resources in 2014. Given the importance placed on hiring at SpaceX, this was a real endorsement of Bjelde. The experiment succeeded, as he still holds the position today.

Bjelde was eager to tell the story of the Falcon 1 rocket for this book, because he wishes every new hire at SpaceX could experience Omelek Island in all of its austerity, that moment when the company had to fly or die.

"It hardened the DNA of the company," he said. "It's still there in decisions we make today. I'll be in a meeting, in our executive conference room, and we've got a picture of Omelek Island, and the Falcon 1 rocket. It recalibrates your mental model. Many of those same old-timers are still here, now in leadership positions. It created the sieve or filter by which we make decisions today. We talk about it all the time. We're always trying to be more efficient, and more like we were in some ways."

Today, he must convince talented young engineers that they won't entirely be throwing away their personal lives by taking a job at SpaceX. Bjelde said he found a balance after the Falcon 1 flight, marrying his college sweetheart in 2010, and starting a family with two young girls.

"It would have been a lot easier for me to stay at JPL, or work someplace like that," he said. "It was a toll and a sacrifice, but I wouldn't change it for the world."

HANS KOENIGSMANN, Vice President of Avionics

Of the very earliest employees Musk hired to work at SpaceX, only Koenigsmann remains. He cherishes his role as one of the older hands at the company, helping to channel the talents of its mostly younger workforce. While many of the brightest engineers have turned their starry-eyed countenances on the Starship program, as vice president of mission assurance, Koenigsmann remains focused on the Falcon 9 rocket and Dragon spacecraft. It is important to get those core programs right. He is stubborn about that, like you'd expect a German engineer to be.

For the last eighteen years, his wife has gone along with his insane work schedule and repeated trips to Kwajalein, Florida, and Texas because she knew he was happiest when working. His children have grown into young adults along the way, finding inspiration in their dad's story.

His youngest daughter, an electrical engineer, recently started work at a small company in Boston. "One of my worries is, maybe she thinks that's how it works," he said. "Maybe she thinks you find a tiny company, and ten years later it's five thousand people. And that's not how it works. I feel like it was the right moment, right time, right people for me."

Koenigsmann credits Musk for much of the company's success. He always made the most difficult decisions. He did not put off problems, but rather tackled the hardest ones first. And he had a vision for how aerospace could be done faster and for less money. From the beginning, Koenigsmann said, Musk wanted SpaceX to build as much of every rocket that it could, to free itself from the cost and schedule vicissitudes of suppliers. But mostly, Koenigsmann said, it was Musk's ability to identify engineering talent and then motivate his employees to do extraordinary things. Musk had a knack for inspiring engineers to do things they believed just beyond their ability, and when they achieved the impossible to reach toward the next goal.

"This company has a lot of talent," Koenigsmann said. "Elon's primary capability is to evaluate people quickly and pick the right people. Yah? He is really good at that. I've been at odds with him on that, too. Sometimes I would interview a guy and say, 'No, he's terrible.' And Elon said, 'No, you take them.' And sometimes the other way around. He was usually right."

GWYNNE SHOTWELL, Vice President of Business Development

After becoming president of SpaceX in 2008, Shotwell did not look back. While much of the aerospace industry often rolls its eyes when Musk makes a prediction about when a key launch will happen, everyone takes Shotwell seriously. More than that, pretty much the whole industry adores her, even though SpaceX has sought to disrupt pretty much everything.

In 2016, after the Falcon 9 rocket's first stage landed on an autonomous drone ship for the first time, the company's competitors might have looked at their own business models with some concern. But they also respected what SpaceX had done. The chief executive of United Launch Alliance, Tory Bruno, sent Shotwell flowers in congratulations.

They have proven to be quite a pair, Elon Musk and Gwynne Shotwell. She understood the industry he wanted to change. And when he pushed for that change, she helped guide him and stood by his side through all of the lawsuits, protests, and pressure campaigns. Along the way, the quality she has come to admire most about Musk is his determined mindset to identify problems and devise solutions.

"He looks at a problem and his reaction isn't *Oh, that's a shame*. His reaction is to go fix it. He's extraordinary," she said. "I've never understood his detractors, the cynics who say he's just in it for the government money. It's such bullshit. He started with Mars Oasis. He wanted to do Mars Oasis because he wanted people to see that life on Mars was doable, and we needed to go there."

Shotwell herself did not believe the Mars stuff in the beginning. "I kind of ignored it," she said. "I wasn't even bought in." She is today.

ELON MUSK, Founder

And what of the ringmaster? Since the early years of his rocket company, Musk has ascended from semi-anonymous dot-com millionaire status to that of multibillionaire, international celebrity. As of this writing he is the fifth-richest person in the world. Yet at his core, Musk remains the same passionate, nerdy, driven person who founded SpaceX to make humans a multiplanetary species. He still speaks with the same earnestness about Mars. Only a goal that seemed preposterous in 2002 now merely seems audacious.

During discussions when I would press Musk to remember details about those first years of his rocket company, he would pause for long moments and close his eyes. In what must have been a measure of his concentration, tiny bits of water welled in his tear ducts. So much has happened since the success of Flight Four. He now leads both a globally dominant rocket business as well as an electric automobile company, Tesla, that seeks nothing less than to migrate humanity from fossil fuels into renewable energy. Musk also started Neuralink in 2016 to build machines that can interface directly with the human brain, and he formed a company to dig tunnels beneath congested cities.

In short, Elon Musk had a lot on his mind when I nudged his

memories back to the tiny island of Omelek. He wanted to help. Musk understands the significance of the Falcon 1 rocket to his life, and how its singular success spurred a transformational change across a number of fronts. Before this book, he had never consented to telling the story in full, or to allow an author free rein inside SpaceX to speak with employees about the company's formative years. But Elon Musk wanted me to talk to everyone for this book. And he meant it.

"It was a high-drama situation," he said of launching rockets from Omelek. "It is a great story. But it is way better in recollection than at the time."

At this Musk laughed, and then took another one of those pauses. His mind moved toward regrets. He had one. Musk did not visit Omelek as much as most other people in this story, but he traveled there enough to know the place intimately. "I remember that island like the back of my hand," he said wistfully. "I think I probably should have chilled out just slightly more. You know, it wouldn't have hurt to have just one cocktail on the damn beach. Just have one. Just go have a drink on the beach with the team. I never did that. It wouldn't have hurt."

It still wouldn't. There is yet time.

ACKNOWLEDGMENTS

I had a blast writing this book. For weeks at a time I mentally escaped into exotic locations half a world away and listened to stories from those who toiled on Kwaj to make the impossible possible. For this experience I have many people to thank.

The list must start with Elon Musk. When I first proposed this book idea in early 2019, he eagerly agreed. His message to me was that I should talk to everyone. With this signal, both current employees at SpaceX and former employees agreed to talk with me at length about their experiences. Elon himself made plenty of time, generously inviting me to sit in on his technical meetings for Starship, Starlink, Raptor, and other projects at the company's factory in Hawthorne. This helped me understand his leadership style. He also opened the doors to his factory-beneath-tents in Boca Chica where a new generation of engineers are building Starship, much in the iterative, fast-paced style of the Falcon 1 days.

I interviewed many dozens of people for this book, and what surprised me most about the earliest employees, people like Tom Mueller,

Chris Thompson, Hans Koenigsmann, Gwynne Shotwell, Tim Buzza, and others is how anxious they were to see this story told. For a lot of them, those hot and sweaty years on Kwaj and elsewhere were both the most difficult and rewarding years of their lives. I can only hope I have honored the considerable trust they placed in me to complete this work.

So many people helped me along the way. This book would not have happened without Jeff Shreve, my agent, who read one of my feature articles on Ars Technica and saw the potential for a full-length book. He eventually convinced me there was one heck of a story to be told about these formative years at SpaceX, and he was not wrong. At William Morrow, executive editor Mauro DiPreta saw the potential of this idea early, and he and Nick Amphlett ably steered me through the writing and editing process. I had never written a book before and I learned so much from them. My day-job editors at Ars, Ken Fisher, Eric Bangeman, and Lee Hutchinson, were also tremendously supportive with flexible schedules for the last eighteen months. At SpaceX, James Gleeson, Verdell Wilson, and Jehn Balajadia were all superb in ensuring that when I needed to talk to people at the company, I was able to do so in a timely manner.

And then there is my family, who put up with me throughout this process and supported my efforts. I inherited a love of writing from my dad, Bruce Berger. He has always been a writer, and when I was young, he would cover my scribblings with detailed edits. My daughters, Analei and Lily, have been wonderful and supportive, offering food, love, and a generous helping of teenage chaff. Finally there is my wife, Amanda. Too many times when she needed me I had my noise-canceling headphones on, or told her I had to write late into the night to finish a chapter. And when I was done with a draft, she would faithfully read it and tell me it was great, even if it was not. I love you; and this book is for you, for always believing in me.

KEY SPACEX EMPLOYEES FROM 2002 TO 2008

ELON MUSK, CHIEF EXECUTIVE OFFICER

Mary Beth Brown, assistant

TOM MUELLER, VICE PRESIDENT OF PROPULSION

Jeremy Hollman, director of propulsion development

Dean Ono, director of in-space propulsion

Glen Nakamoto, Merlin engine designer

Zach Dunn, Merlin development

Kevin Miller, Merlin development

Jon Edwards, Kestrel development engineer

Eric Romo, propulsion analysis

CHRIS THOMPSON, VICE PRESIDENT OF STRUCTURES

Mike Colonno, engineer for primary structures

Florence Li, engineer for primary structures

Chris Hansen, engineer for separation systems

Sam DiMaggio, director of dynamics

Jeff Richichi, director of structures

Rick Cortez, senior structures technician

**HANS KOENIGSMANN, VICE PRESIDENT OF AVIONICS /
CHIEF ENGINEER OF LAUNCH**

Phil Kassouf, senior avionics engineer

Steve Davis, guidance, navigation, and control

Chris Sloan, flight software

Bulent Altan, avionics engineer

Tina Hsu, avionics engineer

Brian Bjelde, avionics engineer

TIM BUZZA, VICE PRESIDENT OF LAUNCH AND TEST

Kenton Lucas, launch ground support equipment

Trip Harris, software

Josh Jung, ground controller

Joe Allen, McGregor lead

Ricky Lim, rocket integration

Anne Chinnery, site development

George "Chip" Bassett, launch infrastructure

Eddie Thomas, senior propulsion technician

GWYNNE SHOTWELL, VICE PRESIDENT OF BUSINESS DEVELOPMENT

David Giger, Flight One mission manager

BOB REAGAN, VICE PRESIDENT OF MACHINING OPERATIONS
BRANDEN SPIKES, CHIEF INFORMATION OFFICER

TIMELINE

2002

MAY 6	SpaceX founded by Elon Musk
OCTOBER 31	First gas generator full-duration test-firing (Mojave, California)

2003

MARCH 11	First Merlin engine thrust chamber firing (McGregor, Texas)
MAY 31	SpaceX employees visit Kwajalein for the first time
JULY 2	First Merlin engine turbopump test (Mojave)
DECEMBER 4	Falcon 1 displayed outside National Air and Space Museum

2004

FEBRUARY 17	First propellant loading of stage one (McGregor)
FEBRUARY 22	First Kestrel engine thrust chamber firing (McGregor)
JULY 1	First complete Merlin engine test-firing (McGregor)
OCTOBER 5	Falcon 1 rocket goes vertical (Vandenberg Air Force Base, California)

2005

MAY 27	Falcon 1 static fire test-firing (Vandenberg)
NOVEMBER 27	First static fire attempt from Kwajalein (Omelek Island)
DECEMBER 20	First Falcon 1 launch attempt (Omelek Island)

2006

MARCH 24	Falcon 1, Flight One (Omelek)
AUGUST 18	SpaceX wins COTS award from NASA

2007

MARCH 21	Falcon 1, Flight Two (Omelek)

2008

AUGUST 3	Falcon 1, Flight Three (Omelek)
SEPTEMBER 3	C-17 carrying Falcon 1 first stage departs Los Angeles
SEPTEMBER 28	Falcon 1, Flight Four (Omelek)
NOVEMBER 22	Falcon 9 full-duration test-firing (McGregor)
DECEMBER 22	SpaceX wins CRS award from NASA

2009

JULY 14	Fifth and final flight of Falcon 1 (Omelek)

2010

JUNE 4	Falcon 9 first launch (Cape Canaveral, Florida)
DECEMBER 8	First launch of Cargo Dragon spacecraft (Cape Canaveral)

2018

FEBRUARY 6	Falcon Heavy first launch (Kennedy Space Center, Florida)

2019

AUGUST 27	Starhopper five-hundred-foot test flight (Boca Chica, Texas)

2020

MAY 30	First astronauts launch aboard Crew Dragon (Kennedy Space Center)
AUGUST 4	First full-sized Starship prototype makes five-hundred-foot test flight (Boca Chica, Texas)

Bulent Altan's Turkish Goulash

"MAKE THIS DISH ANYWHERE A DELICIOUS DINNER IS NECESSARY."

2 to 3 onions

5 to 6 cloves garlic

1 pound ground beef

½ cup (1 stick) butter

Salt and freshly ground black pepper

Three 16-ounce boxes medium shell pasta

One 32-ounce container of plain yogurt

1 tablespoon Turkish red pepper powder, plus more for garnish

Fresh mint leaves

1. Mince the onions finely.

2. Crush the skin of the garlic, peel it, and chop off the woody stem.

3. Break the ground beef into a couple of smaller pieces so it cooks more easily.

4. Melt 1 tablespoon of the butter in a large pot over medium-high heat.

5. Add the onions to the melted butter and stir until they get translucent.

6. Add the ground beef and fry with the onions, making sure to break the ground beef into really small pieces with a sturdy spatula. In the process add salt and black pepper to taste.

7. When the ground beef is cooked through and releases its juices, add the pasta and fill the pot with water until the water is ½ inch (+/−0.100 inch) above the pasta level.

8. In parallel with the pasta operation, mix the yogurt with 2 to 3 tablespoons of salt and crush the previously peeled garlic into a bowl. Mix thoroughly.

9. Place the remainder of the butter in a small pot and add the Turkish red pepper. Don't cook it yet, but keep it ready for when the hungry masses arrive.

10. When the pasta is almost done and has soaked up most of the water, call in the people for dinner.

11. When they have lined up to eat, turn on the small pot and bring the butter to melt and foam up.

12. For each person, assemble the dish by putting the drained pasta–ground beef mixture onto their plate, covering the pasta with the yogurt mixture, then drizzling the butter mixture over that.

13. Sprinkle as much mint and more red pepper as desired on top.

EAT!

This recipe was generously contributed by its creator, Bulent Altan. Now you can bring a taste of the Omelek life into your home.

INDEX